CULTURAL STUDIES IN THE THIRD MILLENNIUM

A CLOSER LOOK AT CULTURAL VALUES

THE CASE OF FRENCH GUESTS AND VIETNAMESE HOSTS

CULTURAL STUDIES IN THE THIRD MILLENNIUM

Additional books and e-books in this series can be found on Nova's website under the Series tab.

CULTURAL STUDIES IN THE THIRD MILLENNIUM

A CLOSER LOOK AT CULTURAL VALUES

THE CASE OF FRENCH GUESTS AND VIETNAMESE HOSTS

THUY-HUONG TRUONG

Copyright © 2020 by Nova Science Publishers, Inc.

All rights reserved. No part of this book may be reproduced, stored in a retrieval system or transmitted in any form or by any means: electronic, electrostatic, magnetic, tape, mechanical photocopying, recording or otherwise without the written permission of the Publisher.

We have partnered with Copyright Clearance Center to make it easy for you to obtain permissions to reuse content from this publication. Simply navigate to this publication's page on Nova's website and locate the "Get Permission" button below the title description. This button is linked directly to the title's permission page on copyright.com. Alternatively, you can visit copyright.com and search by title, ISBN, or ISSN.

For further questions about using the service on copyright.com, please contact:
Copyright Clearance Center
Phone: +1-(978) 750-8400   Fax: +1-(978) 750-4470   E-mail: info@copyright.com.

NOTICE TO THE READER

The Publisher has taken reasonable care in the preparation of this book, but makes no expressed or implied warranty of any kind and assumes no responsibility for any errors or omissions. No liability is assumed for incidental or consequential damages in connection with or arising out of information contained in this book. The Publisher shall not be liable for any special, consequential, or exemplary damages resulting, in whole or in part, from the readers' use of, or reliance upon, this material. Any parts of this book based on government reports are so indicated and copyright is claimed for those parts to the extent applicable to compilations of such works.

Independent verification should be sought for any data, advice or recommendations contained in this book. In addition, no responsibility is assumed by the Publisher for any injury and/or damage to persons or property arising from any methods, products, instructions, ideas or otherwise contained in this publication.

This publication is designed to provide accurate and authoritative information with regard to the subject matter covered herein. It is sold with the clear understanding that the Publisher is not engaged in rendering legal or any other professional services. If legal or any other expert assistance is required, the services of a competent person should be sought. FROM A DECLARATION OF PARTICIPANTS JOINTLY ADOPTED BY A COMMITTEE OF THE AMERICAN BAR ASSOCIATION AND A COMMITTEE OF PUBLISHERS.

Additional color graphics may be available in the e-book version of this book.

Library of Congress Cataloging-in-Publication Data

ISBN: 978-1-53619-209-4

Published by Nova Science Publishers, Inc. † New York

Contents

Preface		vii
Chapter 1	Introduction	1
Chapter 2	Literature Review	13
Chapter 3	Methods	59
Chapter 4	Results	77
Chapter 5	Interpretation and Discussion	97
Chapter 6	Conclusion and Implications	121
References		125
About the Author		151
Index		153

PREFACE

With a history of attracting French tourists since the 19th century and particularly during the colonial period, Vietnam has re-emerged as one of the most popular Asian destinations for the French. A number of factors augur well for a further increase in visitation. As a former colony, Vietnam has nostalgic appeal for many French visitors. After the release of the films L' lndochine, L 'amant, Le Cyclo, Papaye Vert, and the documentary-drama Dien Bien Phu in 1993, France became the main tourism source market for Vietnam.

As the ethnic composition of France has changed, the French are interested in learning about cultures generally and Asian cultures, in particular. The prospect of discovering new destinations and cultures encourages them to travel long-haul. They are quality conscious and culture loving, and prefer to use their own language when traveling overseas, favouring their own cuisine even though they do enjoy the cuisine of the host countries. Travel to Vietnam offers the prospect of fulfilling a number of their preferences. Another attraction for French tourists is that Vietnam is a member of the Francophone Community. Since 1988, the French legacy has experienced a resurgence with the renovation of colonial-style properties and restaurants. Tourists have been impressed by the French-style architecture, accommodation and cuisine that is available.

Despite Vietnam's long history of contact with France, Vietnamese service providers are relatively ignorant about Western countries, their people and their values. Up to now, Vietnam's tourism authorities have paid little attention to the role of cultural understanding in the tourism development process and Vietnam's overseas tourism promotions have given minimal acknowledgement of the cultural characteristics of source markets including those conducted in France. Servicing French tourists is likely to be a challenge for Vietnamese service providers because of the substantial differences between the respective host and guest cultures and rules of behaviour.

International tourism generally involves a cross-cultural component, particularly in the case of encounters between tourists and service providers. If it is accepted that the cultural values of Western travellers are different, it seems reasonable to conclude that the Vietnamese service providers need to consider the effect of this cultural dissimilarity on tourist–host mutual perceptions and social interaction in the intercultural service encounters. An understanding of areas of potential tourist dissatisfaction may assist the service providers to anticipate prospective negative perceptions and to address them, thereby contributing to overall holiday satisfaction, and improve the prospects of repeat visitation.

It is common for destinations to attract visitation from different source countries and cultures. Nonetheless, consumer behaviour literature on cross-cultural perceptions and interaction have been largely limited to homogeneous sample populations from Western countries. Furthermore, until now there are no published studies have specifically examined tourist–host service encounters interaction and mutual perceptions in the context of Vietnam as a holiday destination. Consequently, this research has both practical and academic significances. From the theoretical perspectives, this study provides an augmented comprehension on Argyle's, Rokeach and SERVQUAL models. From the practical standpoints, this study offers service managers and marketers a heightened understanding of cross-cultural awareness for improving customer satisfaction. It acclaims an evidence base that can guide provision to meet the needs of international tourists with particular reference to the impact of rules of behaviour on tourist–host service encounters interaction. It accentuates the effect of cultural backgrounds on tourists' perceptions towards and satisfaction with service quality. By this means, it applies the empirically based models to development related challenges confronting the tourism sector in the cross-cultural settings for designing appropriate strategies with the aim of gaining a competitive advantage.

Keywords: French Guests, Vietnamese Hosts, cultural values, rules of behaviour, interactions, satisfaction, service attributes and performances

Chapter 1

INTRODUCTION

VIETNAM INBOUNDTOURISM AND ITS MARKET POTENTIAL

With its vast population, abundant natural and cultural resources and generally robust economics growth, Asia is one of the world most important regions for tourism and trade (Genkin, 1991). This provides an opportunity for Vietnam's tourism sector to take advantage of wider regional development trends. Various treaties between Vietnam and other ASEAN countries facilitate tourism flows between these nations. Asia's growing middle class is a particular target for Vietnam since it is suitable for those seeking relaxation close to home. Additionally, the strength of Vietnam also lies in its location in relation to neighbouring international gateways. This provides opportunities for the country to develop an intra-regional tourism strategy. According to Pookong and King (1999), all countries in the greater Mekong sub-region are endowed with abundant historical and cultural heritage, combined with a variety of beautiful and unspoiled natural environments. Those qualities combine to make the sub-region an attractive tourism destination.

Located in South East Asia, Vietnam is well place to integrate with wider tourism development trends both regionally and globally. The destination appeals to outsider because of its long history, its culture and its unique customs and habits. The history and development of Vietnam has produced a legacy of grand culture, history and artistic heritage yielding strong tourism attractions. In terms of potential tourist resources, the country is well endowed, and the market potential presents the country with good opportunities for tourism growth (Jansen-Verbeke & Go, 1995). From a natural and cultural perspective, the country has much to offer to holidaymakers. It has beaches, caverns, marine lands and many places of unique and natural beauty including exotic plants and animals. It is also endowed with a rich cultural heritage such as art, architecture, handicraft, customs and habits and tradition of ethnic groups. These characteristics or attributes form the basic potential of a diverse range of tourism products from coastal and beach tourism, through adventure and ecotourism to cultural heritage and urban tourism. Vietnam is therefore a destination appealing to a wide cross-section of travellers.

Vietnam is a developing country where rural overpopulation and the demands of an emerging and diversifying urban economy are rapidly transforming economic and social relationships (Cooper and Hanson, 1997). With the growth of travel worldwide, many underdeveloped countries around such as Vietnam have been able to improve their economies by increasing exports via low-cost production, but also by tourism. Vietnam is the latest Asian country to declare the importance of tourism to National Development (VNAT, 2019). As tourism rapidly expands into Indo-China, Vietnam is trying to position itself to capitalize on this emerging industry.

The formulation of the *Doi Moi* policy brought privatization and Western management practices to the tourism industry and has placed pressure on Vietnam's tourism industry to professionalise in response to rising guest expectations. The opportunity to act independently and the reality of having to satisfy customers or lose business are emerging as powerful motivations for Vietnamese managers. Success is more readily achievable in the larger cities, where the local economy is stronger and visitation base is higher. The establishment of various joint-venture enterprises has signalled the emergence of a cross-cultural style of operation. Evidence exists that the Vietnamese are culturally dissimilar to Westerners but share a number of common characteristics. Such cultural characteristics affect the management of businesses both internally and in relation to others. Over 60 years of Communism in Vietnam has left a legacy of distrust around outsiders including within business. This leads to a tendency to express strongly held opinions individually rather than in-group situations. Face-to-face interactions occur within the context of a top-down or vertical organization structure used by Vietnamese managers which discourages the type of horizontal networking relationships which are commonplace within Western firms.

Since the introduction of *Doi Moi*, Vietnam's economy has faced a difficult transition into a regulated market economy though tourism appears to have faced relatively well through this period. Its success has been attributable to the high expectations of service that international business and leisure travellers have. Inbound travel, foreign trade, and entrepreneurship have all grown. The main challenge has been human resources because of the different cultural values, rules of behaviour, political history, social systems, and business practices prevalent between Vietnam and Western countries. Vietnam's tourism managers have shown a mixture of awareness and ignorance towards the theory and practice of management relative to managers. Front-line employees are removed from the approaching changes. The absence of tourism education and a developed labour market contributes to inexperience and a lack of knowledge on the parts of employees. Vocational training is a particular challenge because of the experience due to years of hard-line government policies and attitudes. A key challenge for tourism service providers is making provision for tourists from different cultural background; to achieve effective management practices, and to be more functioning in the delivery of products and services.

The Vietnamese government's acknowledgement of tourism as a significant economic sector and the increased economic reforms, have affected the development of tourism. The emergence of a free-market economy has created a more appropriate context for the development of a stable tourism industry. Favourable policies toward foreign investment have prompted a range of newly established international hotels thereby providing an increasing supply of rooms and an upgrading of tourism facilities. The continuous improvement to international relations and a gradual relaxation of visa regulations have contributed to increased visitation. This, associated with attractions in terms of geography, economic position and international commodity exchange should create an impetus for tourism development in Vietnam and assist the country to follow with South East Asia's tourism development trends.

Western markets are the other major source of international tourists to Vietnam with nearly 40 per cent market share. These markets will account for the second largest source of international tourists to Vietnam. Inbound tourism from France has also been increasing, though the growth rate has been slower than in the case of Chinese and American markets. The American and French markets are expected to stabilize as the second and third most important markets, despite the recent strong growth from other Western countries such as Australia, Britain, Austria and Germany. VNAT has been increasing its promotional activities in France and the USA with a view to gaining a share of these markets (VNAT, 2019).

Vietnam is becoming increasingly popular with international tour operators both as a standalone destination as well as being linked with tours to other countries. The country is popular with group tours as well as independent travellers and has become a popular destination on the East Asia backpacker route (Millington, 2001). Furthermore, the deflation of the Vietnamese Dong relative to leading international currencies has made Vietnam a more affordable destination. A combination of affordability plus a growth in international marketing by the government has stimulated strong tourism growth.

The 2018 was a successful year for Vietnam tourism. The country was honored by the World Travel Awards as the leading tourist destination and as the best golf tourism destination in Asia, and hailed for its wide variety of tourism products and services. In 2019, Vietnam served over 18 million foreign tourists – the highest number recorded, and 85 million domestic holidaymakers, earned 755 trillion VND (32.6 billion USD) from tourism services, and contributed 9.2 percent of GDP. This year has reflected outstanding achievements of the Vietnamese tourism sector (VNAT, 2019).

Nevertheless, despite the opportunities, Vietnam has many challenges in terms of infrastructure and resources training and management. Due to many years of war and destruction, the country needs to pay urgent attention to maintaining and preserving its national heritage. According to Jansen-Verbeke and Go (1995: 315), Vietnam is not ready yet for a "large influx of tourists," particularly since the country still lacks suitable infrastructure, accommodation facilities, an appropriate tourism organisation and qualified staff to make a smooth change to a market economy tourism sector. Similarly, Theuns (1997) reported that there is an urgent need for upgrading the road infrastructure in order to be able to comply with increased tourist traffic. The level of service also needs to be improved to satisfy the expectations of international travellers and enable it to compete with other countries in the Southeast Asian region, especially Thailand. Cooper (2000) has identified an urgent requirement for basic statistical information on visitor numbers and characteristics in order to obtain effective tourism's policy formulation.

Addressing the challenges of cross-cultural exchanges is important for the tourism industry. Tourism service providers rely on having access to skilled labour that can offer high quality and consistent service. Employee attitudes have however been slipping due to the prevalence of communist ideas with Vietnam's drive towards more open markets having emerged only recently. It is vital to understand how tourists and service providers differ in terms of social or professional interactions, and especially tourists' perceptions towards and satisfaction with service quality in the cross-cultural service encounter.

BILATERAL RELATIONS BETWEEN FRANCE AND VIETNAM

France today is one of the most modern countries in the world and is a leader among European nations. It plays an influential global role as a permanent member of the United Nations Security Council, NATO, the G-7, the G-20, the EU, and other multilateral organizations.

The French economy is diversified across all sectors. The government has partially or fully privatized many large companies, including Air France, France Telecom, Renault, and Thales. With a total population of 67.8 million (July 2020 est.), its purchasing power parity (GDP) is estimated at $2.856 trillion (2017 est.) (US Central Intelligence Agency, 2020).

France carries out a comprehensive foreign policy, acting as a bridge and pioneer in dealing with global issues. The European country also promotes economic diplomacy and boosts French images and products through cultural values, French language, its heritage, tourism and sports. France contributes 10% of the world's total in development aid and ranks fourth in the global donor league table.

France was one of the first Western countries to support Vietnam's reform policy and has been supporting its development and outreach for over 20 years. It has made significant cooperation efforts in terms of official development assistance. France has long been the second leading bilateral donor after Japan, with the Agence Française de Développement contributing nearly €2 billion since 1994. On 25 September 2013, France and Vietnam signed a declaration on strategic partnership aimed at strengthening the relationship in all areas (political, defence, economy, education and culture).

France is the Europe's fifth largest trading partner of Vietnam with bilateral trade of US$4.6 billion in 2017. Two-way trade volume in the first six months of 2018 reached US$2.3 billion. In 2017, France ranked third among European countries and 16th out of 114 countries and territories investing in Vietnam, with 512 valid investment projects posting a total registered capital of US$2.8 billion. Vietnamese businesses have so far invested in nine projects in France with a total investment capital of over US$3 million.

Since the early 1980s, education and training cooperation between the two countries has been established and boosted. France always considers education and training a priority in its cooperation activities in Vietnam, focusing on French language teaching and human resources training at undergraduate and postgraduate level in various fields, including economics, banking, finance, law, and new technology. Vietnamese students in France have increased by 40% over the past 10 years with more than 7,000 Vietnamese students currently studying in the country.

The economic cooperation between Vietnam and France has well developing. In 2017, France is the Europe's fifth largest trading partner of Vietnam with bilateral trade of US$4.6 billion. Two-way trade volume in the first six months of 2018 reached US$2.3 billion. In terms of development cooperation, France is the leading bilateral official development assistance (ODA) donor for Vietnam, with Vietnam ranking second among Asian countries to receive French ODA, with total pledges of up to US$18.4 billion since 1993.

France also ranks seventh among countries and territories investing in the tourism sector in Vietnam, with 14 projects worth a total US$188 million. Vietnam also identifies France as its key tourist market. The bilateral cooperation in the field of science and technology, security-defence, cooperation between localities, and health has left many imprints. As members of the International Organisation of La Francophonie (OIF), the two countries have implemented many cooperation activities within the OIF. In 2019, a programme to promote *"Vietnam as a safe and attractive tourism destination"*. With 15 weekly direct flights connecting Paris with Hanoi and Ho Chi Minh City, France is a potential tourism market of Vietnam, and is a gateway for other European travelers to visit Vietnam (VNAT, 2019).

FRENCH TOURIST MARKET

France has played an important role as an outbound tourism country through its economic contributions to world tourism with 53.32 million overnight trips were taken abroad from France in 2018, up from 28.51 million in 2017. France is the fifth largest outbound market for expenditure in the world (World Tourism Organization, 2018).

Whilst the French have been less dynamic as outbound travellers than their northern European neighbours, outbound tourism has recently emerged and France is now the third largest outbound market in Europe after the UK and Germany in terms of trips, length of stay and expenditure. Unlike many of their European counterparts, most French people travel abroad independently, planning their own holiday agendas and finding their own accommodation. This creates a challenging environment for tour operators and travel agents. Though the French still prefer to book their holidays at least two to four months in advance, last-minute bookings have grown. The use of the Internet is also increasing mainly as an information source. As relatively wealthy country, France is considered to be a quality market since travellers spend more per capita and per trip than most of their neighbours. They spent a total of US$ 18.6 billion on travel in 1999, although this is due more to France's large population than to a strong propensity to travel abroad (Euromonitor, 2002).

The average age of French travellers is the highest in Europe with the 60-plus age group accounting for 30% of all trips. Most overseas trips are made by couples and this group has the highest outbound departure rate. France is the only major travel market where female travellers are in the majority. French long-haul travellers account for more than 50% of all departures. They are well educated, with 36% having completed a college or university degree and their level of income is over FRF 15,000 a month (Jacquemin, 2001). The French favour touring holidays followed by sun and beach and city trips. Holidays account for 64% of trips abroad, with visiting friends and relatives 24% and business travel 12%. The French are also keen on special interest holidays, especially diving, golf, surfing, sailing, and holidays involving other sporting activities (Jacquemin, 2001).

As the ethnic composition of the country of France has changed, the French are interested in learning about cultures generally and Asian cultures, in particular. The prospect of discovering new destinations and cultures encourages them to travel long-haul. They are quality conscious and culture loving, and prefer to use their own language when traveling overseas, favouring their own cuisine even though they do enjoy the cuisine of the host countries (Jacquemin, 2001).

In some special way, Vietnam appeals to several occidental tourism generation destination tourists such as French, Americans and Australians as their forefathers were directly involved as colonial powers or aiding a colonial power against the national forces of Vietnam. War memories are one of the many reasons for these visitors (veterans) to make a pilgrimage to the erstwhile theatres of pitched battles. Alongside this special group of visitors, the occidental urban middle class explorers who evince a keen interest in knowing other cultures and traditions also make a contribution. The senior market is another promising target for Vietnam's tourism sector.

Travel to Vietnam offers the prospect of fulfilling a number of French preferences. Tourists have been impressed by the French-style architecture, accommodation and cuisine that is available, though incorporation of these characteristics within tours declined from 1975 to 1988 when Eastern bloc countries emerged as the major source of tourists to Vietnam. Since 1988, the French legacy has experienced a resurgence with the renovation of colonial-style properties and restaurants. In catering to the French market, inbound tour operators now aim to provide high quality French cuisine and knowledgeable French-speaking guides.

With a history of attracting French tourists since the 19th century and particularly during the colonial period, Vietnam has re-emerged as one of the most popular Asian destinations for the French. A number of factors augur well for a further increase in visitation. As a former colony, Vietnam has nostalgic appeal for many French visitors. After the release of the films *L'Indochine*, *L'amant*, *Le Cyclo*, *Papaye Vert*, and the documentary-drama Dien Bien Phu in 1993, France became main tourism source markets for Vietnam. Another attraction for French tourists is that Vietnam is a member of the Francophone Community.

Over the years, the generally favorable rate of exchange between the Euro and the Vietnamese Dong has reduced the cost of holidays in Vietnam. The Vietnamese tourist sector has acknowledged the importance of the French market with the number of visitors coming just behind China, Japan, Taiwan and the USA. Approximately, 60-70 percent of French visitors came to Vietnam on tour. Given the generally low propensity to travel in France, the market offers greater development than the more mature markets of Germany and the UK, which are closer to reaching a demand ceiling. France's potential for long-haul travel is particularly high and has been estimated at some 9.1 million (Jacquemin, 2001).

Overall, Vietnam attracts visitors from various source markets and such diversity has become commonplace in major tourism receiving areas. The practice of international tourism generally implies cross-cultural experiences on the part of both tourists and service providers. Given the discernable differences between the cultures of Asian and Western; and since European culture are heterogenous, an understanding of the influence of cultural values and rules of behaviour on the expectation and perception of French travellers will be critical for the effective operation of tourism in Vietnam. Awareness of the behavioural differences between French tourists and those of the Vietnamese should assist hosts to understand the expectations and perceptions of tourists. An understanding of any unsatisfactory experiences of French tourists should inform the development of strategies to address negative perceptions and enhance holiday satisfaction. By taking account of cultural values and rules of behavior, and by analysing their influence on the service interactions between hosts and guests, Vietnamese service providers should develop a stronger understanding of the rules and expectation of Western tourists, and be able to respond to the needs of the French tourists.

It is common for destinations to attract visitation from different source countries and cultures. Nonetheless, consumer behaviour literature in evaluating the cross-cultural perceptions and interaction of hosts and guests have been largely limited to homogeneous sample populations from Western countries. Additionally, the influence of cultural differences on service encounters' interaction and perceptions has been relatively under researched. Until now, there are no published studies have specifically examined the tourist–host service encounters interaction and mutual perceptions in Vietnam. As a result, the major objectives of this study are:

- To inspect diverse cultural values and rules of behaviour prevailing amongst French tourists and Vietnamese service providers,
- To examine the effect of cultural values and rules of behaviour on interactions exhibited by French tourists and Vietnamese service providers during the delivery and consumption of services,

- To identify the impact of cultural differences on hosts-guests mutual perceptions and satisfaction towards service attributes and performance, and
- To determine which cultural themes merit particular emphasis in tourism promotional strategies which are targeted at the French tourist market.

Chapter 2

LITERATURE REVIEW

CULTURES VERSUS SUBCULTURES

In line with the title of this book, the key objective of this study is to assess the impact of cultural values and rules of behaviour on hosts-guests inter-perceptions and social interactions in the service encounters. For that reason, this Section highlights various definitions of culture, subcultures, cultural dimensions and values theories. Furthermore, the variances between Eastern and Western cultures are well emphasised in excess of numerous empirical studies from hospitality, tourism and marketing fields.

The term "culture" first entered the anthropological literature in the 1800s drawing upon the German word *Kultur*. Goodenough has described culture as a body of distinct patterns of behaviour transmitted across generations through learning (1981). These shared and learnt patterns embrace all of the things which define human beings. Culture embraces and may be "observed" in terms of both material and non-material elements. The material form of culture refers to productive forces and to everything necessary to support human life (Urriola, 1989) or else to material objects and artifacts (White, 1959). Defining the characteristics of culture is usually expressed through the cultural heritage of particular societies. Such material manifestations may be viewed as cultural expressions, traditions and resources. They are affected by interactions with other cultures and are often the most readily observable aspects of cultural change. Culture may be reflected in material culture, expressions or community traditions (Goodenough, 1981). Traits help to distinguish particular members of group from other members and from outsiders. The material manifestations of culture are useful "cultural markers" distinguishing between, or identifying cultural changes in groups over time.

The non-material or the spiritual, ideational form of culture includes cultural beliefs and values, attitudes, and perceptions. Culture has been an ideological entity comprising values, norms, customs and traditions (Rokeach, 1973) or morality, tradition, and customs (Urriola, 1989). Other commentators have excluded material objects from the concept of culture (Goodenough, 1971). The material manifestation of culture is driven by the ideal elements characterising culture as beliefs, values and norms. These elements influence how world events are perceived, interpreted and expressed through response behaviours. A belief may be viewed as a symbolic statement about reality. Beliefs also exist in the personal domain, but it is the collection of shared beliefs that define cultures. Values are symbolic statements of what is right and important and help to define propriety within a culture. Norms are symbolic statements of expected behaviour which define the limits of acceptable behaviour, especially for community members. The current study focuses on the ideological and material aspects of French and Vietnamese cultures.

Culture is broad and multidimensional and difficult to define (Edelstein et al., 1989). Over 160 definitions of culture have been proposed in the literature (Kroeber and Kluckhohn, 1985). Culture has been defined and conceptualized in diverse social science and business related disciplines including anthropology, sociology, psychology, intercultural communication, marketing and management. These definitions range from culture as "everything" to narrowly defined views. Porter and Samovar described culture as ".... the form and pattern for living. People learn to think, feel, believe, and strive for what their culture considers proper. Language habits, friendships, eating habits, communication practices, social acts, economic and political activities, and technology, all follow the patterns of culture" (1988, p. 19). Similarly, Assael (1992) and Mowen (1993) have interpreted culture as a combination of ideological and material elements such as what and how people eat, what they wear and what they use. Tse et al. in Mowen has characterised cultures on the basis of their "degree of regulation of behaviour, the attitudes of the people, the value of the people, the lifestyle of the people, and the degree of tolerance of other cultures" (1988, p. 581). Berthon (1993) has conceptualised culture as the result of human actions and has shown the link between "mental programming" and its behavioural consequences. Herbig (1998) defined culture as "the sum of a way of life, including expected behaviour, beliefs, values, language and living practices shared by members of a society, consisting of both explicit and implicit rules through which experience is interpreted" (p. 11). Pizam referred to culture as "an umbrella word that encompasses a whole set of implicitly, widely shared beliefs, traditions, values, and expectations that characterizes a particular group of people" (1999, p. 393).

Potter (1989) reported that the extent to which people share meanings depends on their awareness of their own values and beliefs and of the values and beliefs of others. Once they become aware of the differences between these beliefs and values, they can adjust their behaviour with a view to enhancing their capacity to work successfully with people from other cultures. Hawkins et al. (1998) highlighted the completion of culture as a concept which encapsulates most of the influences on an individual's thought processes and behaviour. Herbig (1998) indicated that cultural beliefs, values and customs are followed as long as they yield satisfaction. For this reason, culture is constantly evolving to meet the needs of society and is modified or replaced if a specific standard of conduct does not fully satisfy the members of a society. Nieto (1996) defined culture as:

> "the ever-changing values, traditions, social and political relationships, and worldview created and shared by a group of people bound together by a combination of factors that can include a common history, geographic location, language, social class, and/or religion, and how these are transformed by those who share them. Thus, it includes not only tangibles such as foods, holidays, dress, and artistic expression, but

also less tangible manifestations such as a communication style, attitudes, values and family relationships." (Nieto, 1996, p. 39)

Potter (1989) and Wallerstein (1990) have noted how different groups of people do things differently and perceive the world differently. According to Triandis (1972), there would be no cultures if there were no differences. Hofstede's study (1980) provided evidence of differences and similarities between cultures. Similar to Triandis (1977a, 1977b), Landis and Brislin (1983) reported that cultural differences cause different interactional behaviours, misunderstandings of interpretation, and even conflict. In cross-cultural contact the cultural differences tend to reduce interaction amongst members of different cultures (Albert and Triandis, 1979). This indicates that the analysis of interactional behaviour and its interpretation is critical for the analysis of cross-cultural perceptions and satisfaction. For the purpose of the present research, the findings of Triandis (1972, 1977a, 1977b, 1979), Triandis et al. (1972a, 1972b) and Landis and Brislin (1983) provided a foundation for the development of hypotheses on the basis of the cultural differences between tourists and hosts will decrease on the service encounter interactions and will impact on overall perceptions and levels of satisfaction.

There are many other definitions of culture. Most studies refer to culture in terms of psychological concepts such as values, norms, rules, behaviour, perceptions, attitudes, beliefs, symbols, knowledge, ideas, meanings and thoughts (Bennett and Kassarjian, 1972; Argyles 1978; Peterson, 1979; Leighton, 1981; Camilleri, 1985; Ember and Ember, 1985; Mill and Morrison, 1985; Moutinho, 1987; Robinson and Nemetz, 1988; Kim and Gudykunst, 1988). A range of these definitions has been used in the present study to analyse the cultures of Vietnamese hosts and French tourists. Particularly attention is given to Kroeber and Kluckhohn's (1952) comprehensive view of culture. This widely accepted definition was reconfirmed by Adler (1997) as:

> "Culture consists of patterns, explicit and implicit, of and for behaviour acquired and transmitted by symbols, constituting the distinctive

achievement of human groups, including their embodiment in artifacts; the essential core of culture consists of traditional (i.e., historically derived and selected) ideas and especially their attached values; culture systems may, on the one hand, be considered as products of action, on the other, as conditioning elements of future action." (Kroeber and Kluckhohn, 1952, p. 181)

Kroeber and Kluckhohn's (1952) definition of culture has provided a useful basis for the objectives of the present study as it refers to culture as values, rules of social behaviour, perceptions, and differences and similarities between people. This definition was chosen because cross-cultural satisfaction of tourists and service providers appears to be influenced by the differences and similarities between tourist and host cultural values, rules of behaviour and perceptions.

Subculture may be viewed as an attribute of a collectivity which may be variously conceptualised as the human race, a nation-state, a society, a community, an industry, a profession, a corporation or an organization. On this basis, culture may be invested at various levels using different units of analysis such as societal culture, industrial culture, and corporate culture. Some collectivities may be viewed as subsets of others and in these cases it may be expected that higher level culture constrains but does not determine subcultures (Terpstra and David, 1985). Lower level cultures would be concerned with more specific situations. Five levels of culture may be distinguished analytically for the purposes of studying management practices: universal (human) culture, societal/national culture, business culture, industrial culture, and corporate (organizational) culture (Ko et al., 1990; Terpstra and David, 1985). The current study focuses on the universal (human) culture and the societal and national levels of culture.

According to Kluckhohn and Strodtbeck (1961), dominant and variant cultures are clearly distinguishable. A distinction is also evident between public and private cultures (Goodenough, 1971). Dominant cultures may be viewed as collections of subcultures which are based on geographic regional, race, ethnicity, social and economic class. Schneider and Barsoux (1997) have indicated that regional subcultures evolve due to differences in geography, history, political and economic forces, language and religion.

Horowitz reported that ethnic differences are distinguishable on the basis of "colour, language, religion, or some other attribute of common origin" (1985, p. 41). Each subculture exhibits behavioural characteristic which distinguishes which it from others within a parent culture. Subcultures provide their members with norms and rules, which guide behaviour, interactions and thinking within these subcultures. Subcultures provide their members with a different set of values and expectations and interactions as a result of regional differences, yet both groups share the patterns of the dominant culture. The minor variant subcultures therefore differ from the major dominant culture. A dominant culture is indicative of the prevailing forms of public social interactions. The variant minor subcultures indicate that the forms of private social dominant culture may not be easy to grasp because in reality they may be members of various subcultures. Their backgrounds may differ so substantially that they may be incapable of relating appropriately.

In the current study, reference is made primarily to the dominant cultures. It does not consider tourists and hosts subcultures since it is viewed as excessively complex to categorize people into cultural groups and label them on the basis of characteristics. The study focuses on the dominant cultures of tourists and hosts populations from Vietnam and France. It investigates how cultural differences impact on tourist–host mutual perceptions and social interactions in tourism service encounters.

CLASSIFICATION AND MEASUREMENT OF CULTURAL VALUES

As was the case with culture, there are many ways to define values. Being inherently interdisciplinary both concepts have been used for diverse purposes in diverse contexts. In general, values are a core concept of what is desirable within every individual and society. Values have been defined as a system and core of meaning (Sapir, 1949, 1964; Kluckhohn, 1956); a specific preference and belief about these preferences (Catton, 1959; Baier, 1969); individuals' attributes (Barton, 1969), and "attributes of people, as having affective, cognitive and conative elements" (Kluckhohn, 1951, p. 395). Triandis (1972) and Williams (1979) supported Kluckhohn's definition by noting that values are abstract categories and preferences for action exhibiting strong affective components. Kluckhohn (1951) emphasised that it would be impossible to measure order in social life without values. Social values are general principles, which define life situations, their selection and decision-making and can predict social life. He defined values as "a conception explicit or implicit, distinctive of an individual or characteristic of a group, of the desirable which influences the selection from available modes, means and ends of actions". Albert (1968) criticized Kluckhohn's definition for being ambiguous and confusing; however it was supported by Rokeach (1973) and by Adler (1997). Rokeach (1973) refers to values as "an enduring belief that a specific mode of conduct or end-state of existence is personally or socially preferable to an opposite or converse mode of conductor end-state of existence" (p. 5). He agreed with Kluckhohn (1951), viewing values as "means and ends" and noting that values are socially shared, and are conceptions of the desirable and preferences for actions and have strong affective components. He indicated that values are abstract ideas representing aspects of an ideal existence such as freedom, equality, happiness, security, and salvation.

As was suggested by Maslow (1954) in his well-known hierarchical theory of motivations, values which serve adjustive, ego-defensive, knowledge and self-actualization functions may be ordered along a continuum ranging from lower- to higher-order. Smith (1969) has stated that values have an important role in understanding evaluations such as judging, praising or condemning. Values have been important variables in the study of various topics about human behaviour, based on the underlying belief that they guide actions and judgments across specific situations and stimuli. Values are standards which guide ongoing activities and value systems, forming general plans for the resolution of conflict and for decision making. Values lead us to take particular positions on social issues and predispose us to favour one political or religious ideology over another. Rokeach (1979a, 1979b) has further referred to values as "beliefs about desirable goals and modes of conduct" (p. 41) such as how to behave with justice, compassion, humility, sincerity, respect, honour and loyalty or to seek truth and beauty. He emphasized that a value is a system of criteria by which behaviour is evaluated and sanctions applied, a system of social guidelines showing the cultural norms of a society and specifying the ways in which people should behave, a system of standards that permit individuals to decide about relationships with self, with others and with society. Rokeach's (1979a) definition was further emphasized by Feather (1994) who pointed out that values serve to "provide standards or criteria that can be used to evaluate actions or outcomes, to justify opinions and conduct, to plan and guide behaviour."

Lustig (1988) reported that values are "... predictable behaviour patterns, which are stable over time and which lead to roughly similar behaviours across similar situations, are based upon a form of mental programming called values" (p. 56). Being more central to an individual's cognitive system, values are more stable over time (Kamakura and Mazzon, 1991; Kamakura and Novak, 1992). Values are better predictors of an individual's behaviour because of their centrality and stability (Madrigal and Kahle, 1994). Based on these arguments, a value system is relatively stable over time. However, Rokeach (1973) reported that value priorities may be rearranged in the long-term as culture changes. Changes in value affect the thoughts, beliefs, attitudes, actions and social behaviour. Conversely these changes depend on the intensity with which persons within a culture hold these values since value systems may be in conflict; cultural differences in value orientations may lead to disagreements. A person may feel a conflict at an individual level about being polite versus being dishonest. This is similar to the members of a group disagreeing about the importance of values at a societal level. Value differences may jeopardize successful interaction when two people are from different cultures. Samovar and Porter (1988) mentioned that violations of expectations based upon a value system can produce hurt, insult and general dissatisfaction.

In conclusion, the concept of value is not readily definable. As indicated by Adler (1956) the concept of values is vague, lacking real meaning and broad. Values are socially shared standards of behaviour, beliefs about desirable behaviour and preferable existence. Values are patterns of choice that guide persons and groups to prefer certain situations, states of affairs or behaviours towards others. Values influence evaluation of others and ourselves, and they affect perceptions. Samovar and Porter (1988, p. 25) pointed out that cultural values also specify which behaviours "are important and which should be avoided" within a culture. Cultural values influence rules of social behaviour as they are the evaluative aspects and an umbrella concept, that includes such elements as shared values, beliefs, rule, attitude and norms that collectively distinguish a particular group of people from others. Values can be prioritised on the basis of their significance and the conflict of values in intercultural interactions can create clashes.

Values, Culture and Context

There is a relationship between values and culture. Values are both the core of culture (Kroeber and Kluckhohn, 1952) and are dependent on culture (Fridgen, 1991). Culture is rooted in values, and both culture and values held by its members are related (Hofstede, 1980) because culture is a system of shared values of its members (Bailey, 1991). Rokeach (1973) reported that "... differences between cultures ... concern differences in underlying values and value systems" (p. 26). By providing examples of differences between groupings, Rokeach concluded that values differentiate significantly amongst cultural groups. He reported that values indicate cultural differences in thinking, acting, perceiving, understanding of attitudes, motivations and human needs (Rokeach, 1973). In a similar way, Chamberlain (1985) noted that different values are evident between differing cultural groups and, therefore help to differentiate between cultural groups. Segall (1979) also indicated that people from different cultures exhibit different cultural values. However, Williams (1979) argued that while some values are universal, societies exhibit distinct patterns of cultural values because with values "all continuing human groupings develop normative orientations-conceptions of preferred and obligatory conduct and of desirable and undesirable states of affairs ...". The most important types of normative elements are norms (specific obligatory demands, claims, expectations, rules) and values (the criteria of desirability)" (p. 15). Such differences involve not only differences in the relative importance of particular values, but also differences in the extent to which members of each society adhere to particular values, differences in the degree to which the values are universally accepted within a society, and differences in the emphasis which each society attaches to particular values. Because humans are born into and raised within an existing culture, researchers posit that personal values are derived from and modified by, several cultural institutions such as family, work, school, religious institutions, and social organizations (Clawson and Vinson 1978, p. 401). One example is how respect for the elderly is initially learnt and subsequently reinforced in many Eastern and Middle-Eastern cultures. A child who observes parents and significant others treating older people respectfully tends to behave similarly, especially when that behaviour is encouraged by rewards (verbal complements and positive facial gestures) and deviation leads to punishment (frowning, deprivation of valued possessions or privileges).

Value may be viewed differently to value orientation. Kluckhohn and Strodtbeck (1961) indicated that value orientation is a "complex but definitely patterned-rank ordered principle, ... which gives ... direction to the ... human acts ... the solution of common human problems" and they "influence the full range of human behaviour"' (p. 4). The human problems for which people in all cultures must find solutions are how to relate to human nature, nature, activity, time and to other people. Variations in value orientations are the most vital category of cultural variations and, therefore, the fundamental structural feature of culture. Zavalloni (1980) indicated that value orientation is an important variable in the comparison of cultures, because different cultures find different solutions to human problems. The analysis of cultural value orientations can determine similar and different cultural backgrounds (Hofstede, 1980). The major orientations along which cultures differ have been identified in studies by Parson (1951), Kluckhohn and Strodtbeck (1961), Stewart (1971), Hall (1976); Hofstede (1980), Argyle, (1986), Trompenaars (1984, 1993), Hampden-Turner and Trompenaars (1993); Maznevski (1994).

Personal values and cultural values are related. Personal values are an organized system of centrally held beliefs based on what individuals consider to be important at various stages of their lives. Essentially, human values mainly evolve during in the first five years or so of life, and are stored in long-term memory in a hierarchical cognitive structure. Humans acquire values because they serve important cognitive functions. As suggested by Kahle's (1996) social adaptation theory, values help humans to become efficient when dealing with their social and physical environments. While both are called "values," personal and cultural values differ in a number of characteristics. Whereas personal values represent what is important and desirable to an individual, cultural values are promoted by society as preferred and useful for the common good. These values have been accepted and carried from one generation to the next. Individuals are expected to abide by cultural values since they are realities that must be respected (i.e., cultural norms), yet personal values are desired modes of behaviour and an end-state of existence that reflect the needs and objectives in life for an individual. Another difference is that unlike personal values, cultural values generally lack a specific rank order of importance, and are shared by members of a culture with varying intensity. Since values may be applied to individuals (personal values) and groups (cultural values) (Kluckhohn, 1951) that mutually influence each other (Barth, 1966; Meissner, 1971), cultural values may be viewed as main beliefs (principles) or a core around which personal values develop. By examining personal values, it is possible to analyse the cultural values of a particular society. The dominant cultural values kept by society however need not be identical or even similar to individual personal values.

In conclusion, values create culture and distinguish people within or between cultures. Values are essential variables for a comparison of cultures. For the purposes of the current study, values are treated as one of the elements of culture, and the research will examine whether differences in values will result in differences between French tourists and Vietnamese service providers. The concept of cultural value that is used in the present study is similar to Rokeach's (1973) concept of value and Kluckhohn's (1951) value orientation.

According to Maslow (1959); Kluckhohn (1951); Allport (1961); Campbell (1963); Williams (1968); Stewart (1972); Rokeach (1973); Zavalloni (1980); Moutinho (1987), there is a relationship between values and other concepts such as behaviour, attitudes, perceptions, beliefs, rules, norms, interests, motivations, or needs. Values involve an individual's behaviour (Kluckhohn, 1951; Rokeach, 1973) because they are cultural determinants of behaviour (Zavalloni, 1980). Samovar and Porter (1988) have indicated that values prescribe behaviours expected by members of a culture. Peterson (1979) and Fridgen (1991) reported that values guide and rank behaviour. They specified which behaviours are important and which should be avoided within a culture. Lustig (1988) proposed that "values are inside people, in their minds. They are a way of thinking about the world, of orienteering oneself to it; therefore values are mental programs that govern specific behaviour choices" (p. 56). Lustig's definition was supported by Feather (1994) who indicated that "values influence many aspects of our lives, affecting both the way we construe and evaluate situations and the actions that we take in pursuit of important goals." Values exist at a transcendental level to attitudes; thereby values guide attitudes, which in turn guide behaviour. This hierarchical system of Values-Attitudes-Behaviour has also been proven in research conducted by Homer and Kahle (1988) who stated that because values are the most abstract social cognitions, they reflect the most basic adaptive characteristics that serve as prototypes from which attitudes and behaviour are produced. Most people follow normative values that guide their behaviour because failure to do so may be met with sanctions. Similar values predispose a similar way of behaviour, whereas different values reflect behavioural differences (Rokeach, 1973). On this basis, values precede behaviour.

Samovar and Porter (1988) have noted that values are related to attitudes because they both determine attitudes and contribute to the development and content of attitudes (Rokeach, 1973). Attitudes are learned within a cultural context and tend to respond to value orientations in a consistent manner. For example, valuing harmony is indicative of an attitude toward people and towards the nature of relationships between people. Similarity in terminal values determines harmonious interpersonal interactions (Sikula, 1970).

Madrigal and Kahle (1994) have proposed that values are representative of more abstract ideals, positive or negative, that are not tied to any specific object or situation; whereas an attitude is usually directed at specific objects or situations.

Rokeach (1973) further indicated that values differ from attitudes, in their generality or abstractness and in their hierarchical ordering by importance. Values are correlated to attitudes, opinions and lifestyles. They are enduring and end-states that when realized or actualised within a specific cultural context, provide a basis for more specific attitudes and behaviour. They are the conceptual tools and weapons that we all employ to maintain and enhance our self-esteem. All of a person's attitudes may be regarded as value-expressive, and all of a person's values as maintaining and enhancing self-esteem by helping a person adjust to his or her society, defend his or her ego against threat, and test reality. Members of society use values to judge concrete rules, goals or actions. Values are often regarded as absolute although their formation and apprehension evolve in the normal process of social interaction (Theodorson, 1969). A people will draw upon several values concerning a desirable behaviour, for fewer values than in the case of attitudes as there are many as attitudes as encounters. Allport (1961) indicated that values determine attitudes and that value is a more dynamic concept than attitude. Values provide more information about persons, groups and cultures than attitudes (Rokeach, 1968a and 1968b). Values are generally more useful than attitudes for understanding behaviour. Although Campbell (1963) argued that value and attitude concepts are similar, and Newcomb et al. (1965) viewed values as special cases of attitudes and are superior to attitudes. Generally the literature is in agreement that values are superior to attitudes. Bailey (1991) also agreed with Kluckhohn (1951) that "values are individual attributes that can affect such things as the attitudes, perceptions, needs and motivations of people" (p. 78). Since attitudes influence perceptions (Bochner, 1982), values also determine perceptions (Samovar and Porter, 1988). Therefore, the concept of value is also superior to the concept of perception. As values vary from one culture to another, behaviour, rules, attitudes and perceptions also differ across various cultures. Furthermore, Madrigal and Kahle (1994) indicated that the differences between values and attitudes allow for a clustering of societies and market segmentation.

Values can be used as dependent or as independent variables. On the dependent side, they are a result of all the cultural, institutional, and personal forces acting upon a person through his or her lifetime. Similar values are widely held by most members of a culture or subculture. Values are derived from and modified through personal, social and cultural learning. On the independent side, they have far-reaching effects on virtually all areas of human behaviour that are worth investigating and understanding from a social scientific perspective. The major role played by values is as a standard that individuals can use in formulating attitudes and guiding their own behaviour (Clawson and Vinson, 1978). In summary, cultural values are related to the concepts of behaviour, attitudes, rules and perceptions but are superior to them. The value concept will be used as the major dominant cultural variable in the present study to differentiate cultural differences between French tourists and Vietnamese hosts.

The most popular types of values are Kluckhohn's (1956, 1959) and Kluckhohn and Strodtbeck's (1961). They classified five basic value orientations including human nature, nature, activity, time and other people. Hofstede (1991) reported four value dimensions along which various cultures differ such as Power-Distance, Uncertainty-Avoidance, Individualism-Collectivism and Masculinity-Femininity. Schwartz (1994) identified ten distinct motivational value types that are likely to be recognized within and across cultures and used to form value priorities including 1/ Self-Direction, 2/ Stimulation, 3/ Hedonism, 4/ Achievement, 5/ Power, 6/ Security, 7/ Conformity, 8/Tradition, 9/ Benevolence and 10/ Universalism. Furthermore, they depend upon the culture in which the person resides.

Classification and Measurement of Values

Cultural values differ in various ways. Any description of value differences requires a listing of the ways in which cultures could potentially differ. Various researchers have classified the various types of values including Parsons and Shils (1951); White (1951); Parsons (1951, 1953); Albert (1956); Allport et al. (1960). Values may be classified on the basis of their importance within a society. Samovar and Porter (1988) classified values into primary, secondary and tertiary categories. Primary values are the most significant and at the top of the value hierarchy. They specify what is worth sacrificing in human life. Secondary values are very significant too, but not sufficiently strong to mean sacrificing human life. Tertiary values are at the bottom of the value hierarchy. The best-known classifications of values are that of Kluckhohn and Strodtbeck (1961) and Hofstede (1980). Their value categorizations achieved acceptance as dimensions that best distinguish different cultural groups.

Table 1. Measurement techniques of values

Author	Year
Allport-Vernon Values Scale	Allport et al., 1951
Ways to Live Test	Morris, 1956
Allport-Vernon-Lindzey Values Scale	Allport et al., 1960
Survey of Interpersonal Values	Gordon, 1960
Description of Value Orientations	Kluckhohn & Strodtbeck, 1961
Personal Value Scales	Scott, 1965
Questionnaire Measures and Procedures	Mirels & Garrett, 1971; Scott, 1965
Ranking Procedures	Kohn, 1969; Rokeach, 1973
Antecedent Consequent Procedure	Triandis et al., 1972a
Value Survey of Rokeach	Rokeach, 1973
Ideographic Procedures	Zavalloni, 1980
Chinese Value Survey	Chinese Culture Connection, 1987
Chinese Cultural Value Scale	Yau, 1994

Direct and indirect methods have been used to measure values in cross-cultural studies. Direct measurement involves survey research when respondents are asked to rank values according to their importance (Rokeach, 1973), or rate them on a Likert scale (Millbraith, 1980; Mourn, 1980). Indirect measurement is when values are measured indirectly by asking respondents about their self-description, their ideological statements or by describing third persons since perceptions of third persons are influenced by the values of the respondent (Hofstede, 1980). Values can also be assessed through the use of open-ended questions or by observing choices (Williams, 1978). Numerous measurement techniques have been used in cross-cultural studies as described in Table 1. However, not all of these techniques have gained universal acceptance.

There have been several instruments that have been devised to quantitatively measure values. One such instrument is the Rokeach Value Survey (RVS) developed by Rokeach (1973). This is probably one of the most famous and common value measurement instruments (Kamakura and Novak, 1992). Rokeach (1968b, 1971, 1973, 1979a and 1979b) has suggested that cultures develop into instrumental values and terminal values. His definition (1973) was supported by Lovejoy (1950), Kluckhohn (1951) and Kluckhohn and Strodtbeck (1961). They agreed that the total number of terminal values that a person possesses is approximately with about 18 instrumental values. He indicated that attitudes and behaviour depend on whether personal or social values are the priority. The number of values is limited by a person's biological and social make-up and his needs. Values have strong motivational components as well as cognitive, affective, and behavioural components.

From the perspective of conceiving values as modes of conduct or end-states of existence, Rokeach (1973) defined two kinds of values: (a) terminal values, and (b) instrumental values. Terminal values refer to end-states of existence. Values of this type are social or personal, intrapersonal or interpersonal values. They are motivating because they represent the supergoals beyond immediate, biologically urgent goals. Instrumental values are modes of conduct that relate to morality or competence. Instrumental values are motivating because their idealized modes of behaviour are perceived to be instrumental to the attainment of desired end-goals. If we behave in all the ways prescribed by our instrumental values, we will be rewarded with all the end-states specified by our terminal values. The former include freedom, a comfortable life, wisdom, a world at peace and true friendship, whilst the latter include honesty, love, obedience, ambitiousness and independence. Table 2 outlines Rokeach's (1973) value survey with two lists of 18 alphabetically arranged instrumental and terminal values. For the purpose of the present study, the Rokeach Value Survey (1973) will be used to measure the cultural differences between the French tourists and Vietnamese hosts.

Table 2. Rokeach's terminal and instrumental values

Terminal Values	Instrumental Values
A comfortable life *(a prosperous life)*	Ambitious *(hard working, aspiring)*
An exiting life *(a stimulating, active life)*	Brad-minded *(open-minded)*
A sense of accomplishment *(contribution)*	Capable *(competent, effective)*
A world at peace *(free of war and conflict)*	Cheerful *(lighthearted, joyful)*
A world of beauty *(beauty of nature, arts)*	Clean *(neat, tidy)*
Equality *(brotherhood, equal opportunity)*	Courageous *(standing up for one's beliefs)*
Family security *(taking care of loved ones)*	Forgiving *(willing to pardon others)*
Freedom *(independence, free choice)*	Helpful *working for the welfare of others)*
Happiness *(contentedness)*	Honest *(sincere, truthful)*
Inner harmony *(freedom from inner conflict)*	Imaginative *(daring, creative)*
Mature love *(sexual and spiritual intimacy)*	Independent *(self-reliant, self-sufficient)*
National security *(protection from attack)*	Intellectual *(intelligent, reflective)*
Pleasure *(an enjoyable leisurely life, fun)*	Logical *(consistent, rational)*
Salvation *(saved, eternal life)*	Loving *(affectionate, tender)*
Self-respect *(self-esteem)*	Obedient *(dutiful, respectful)*
Social recognition *(respect, admiration)*	Polite *(courteous, well-mannered, kind)*
True friendship *(close companionship)*	Responsible *(dependable, reliable)*
Wisdom *(knowledge, understanding of life)*	Self-controlled *(restrained, self-disciplined)*

A Review of Cultural Dimensions

In the first half of the twentieth century, social anthropologists were convinced that all societies, modern or traditional, face the same basic problems; it is only the answers which are different. These problems and their answers represent "dimensions" of culture. Thus, a cultural dimension is an aspect of culture that can be measured relative to other cultures (Hofstede, 1997). As was noted in the previous sections, many elements differentiate cultural groups based on the examination of cultural differences. There are also a number of dimensions distinguishing between cultures (Parsons and Shils, 1951; Cattell, 1953; Mead, 1967; Inkeles and Levinson, 1969; Ackoff and Emery, 1972; Douglas, 1973, 1978). The most frequently used dimensions are from Parsons' (1951) pattern variables; Kluckhohn and Strodtbeck's (1961) value orientations; Stewart's (1971) cultural patterns; Hall's (1960, 1966, 1973, 1976, 1977, 1983); Hall and Hall's (1987) cultural differentiation; Bond (1987) and Chinese Culture Connection (1987); Hofstede's (1980, 1984, 1991, 2001, 2005) dimensions of cultural variability; Trompenaars (1984,1993,1997); Hampden-Turner and Trompenaars (1993) and Maznevski's (1994) cultural dimensions. According to Gudykunst et al. (1988b) these dimensions have helped us to understand how people behave and communicate differently across cultures and how they deal with social life and human relationships. Amongst the various concepts, the ones by Hofstede (1980), Kluckhohn and Strodtbeck (1961) and Hall (1976, 1977 and 1983) have the major effect upon dealing with social life and human relationships. These dimensions affect the rules of social interaction, the difficulties that individuals experience in relating to others, and individual perceptions (Gudykunst et al., 1988b). As a result of these underlying value dimensions, those within a culture share common beliefs, attitudes, customs, meanings and behavioural norms (1972, 1979).

TOWARDS AN INTERCULTURAL UNDERSTANDING

Cultural differences manifest themselves in many ways. Triandis (1972, 1979, 1990) has indicated that there are major differences across cultures. Part of the explanation for cultural differences between the East and West is thought to be a result of the differences in respective cultural orientations and perspectives regarding the society-member relationship. Various scholars have identified a dichotomy between Collectivism (East) versus Individualism (West). The Collectivism-Individualism paradigm provides a powerful explanatory framework for differences in cultural values, human behaviour, and interpersonal communication (Gudykunst et al., 1988). In individualistic cultures, "people are supposed to look after themselves and their immediate family only" whereas in collectivist cultures, "people belong to in groups or collectivities which are supposed to look after them in exchange for loyalty" (Hofstede and Bond, 1984, p. 419). Kim and

Gudykunst (1988) reported that cultures vary in important ways across societies in terms of values, norms, beliefs, communication styles and many other assumptions. Scollon and Scollon (1995) identified numerous aspects of culture that are significant for the understanding of cultural differences. Hofstede (1991); Czinkota and Ronkainen (1993) and Trompenaars (1993) suggested a range of elements that generate cultural differences. The differences across cultures may be observed in social categories such as class, status, role, hierarchy, in the relationships between individuals, in the attitudes towards nature, human, time, activity, and many others. They are reflected in different patterns of verbal (language and paralanguage such as laughing, crying, questioning), and non-verbal communications (such as body language, facial expressions, eye contact, head movements, gestures, use of time and space). They can be found in taking initiatives or responses, in saving face, in perceiving sense of shame, in feelings of obligations or responsibility, in avoiding embarrassment and external appearance and even in touching, in looking or in standing (Hall, 1955, 1959, 1976, 1983; Argyle, 1967, 1978; Taylor, 1974; Gudykunst and Kim, 1984a; Damen, 1987; Dodd, 1987; Thiederman, 1989).

According to Argyle (1967) and Triandis (1972, 1979) there are different rules of social behaviour, different techniques of establishing and preserving relations and different interaction patterns such as greetings, self-presentation in every culture. Similarly, these differences are also demonstrated in how a conversation is begun by the degree of expressiveness, showing emotions, frankness, intensity, persistency, intimacy as well as the volume of interaction (Jensen, 1970) or in the ways of defining interpersonal relations and attributing importance to social interactions (Wagatsuma and Rosett, 1986). There are differences in addressing people, showing warmth, apologizing, joking, gift giving, farewelling, asking personal questions, complimenting and complaining, expressing negative opinions and expressing like or dislike. For instance, Nomura and Bamlund (1983) noted that there are cross-cultural differences in expressing dissatisfaction and criticism. Similarly, Argyle (1978) reported that these differences also exhibited in the reasons and opinions, exaggerations, and moral rules about telling the truth. Furthermore, cultural differences can cause social interaction problems which lead to different perceptions between participants from different cultural backgrounds. For example, different patterns of verbal and non-verbal communications may create serious mistakes and lead to confusion, misinterpretation and misunderstandings (Argyle, 1967) and have an effect on the perceptions of others (Jensen, 1970; Wolfgang, 1979 and Samovar et al., 1981). According to Argyle (1967), assuming that the contact participants are culturally the same or similar, they may reject each other if they do not conform to each other's cultural patterns of interaction and expected standards. In summary, there are many differences between cultures. These lead to interactional difficulties and influence patterns of social interaction, perception and satisfaction. Therefore, it is very critical to analyse cultural differences and their impact on interactional behaviour, on the levels of inter-perception and tourists and host satisfaction.

Cultural Differences between Eastern and Western Societies

The review of major sets of cultural orientations also demonstrated an evidence of many empirical and non-empirical studies concerning the differences in cultural values between Eastern and Western societies. Within the Western societies, there are also great differences between European and Anglo-Saxon cultures. Even inside the European countries, the value orientations are quite different from each other. Hofstede (1996) noted that Europeans are different in their mental programming and the differences are quite large. Lessem and Neubauer (1994) attempted to characterize the European values and philosophies that dominate European culture. They determined the four most significant European values as pragmatism, rationalism, holism, and humanism. The Latin countries such as France, Belgium, Italy, Portugal, Spain and all the countries speaking the Roman language, plus Greece, are characterized by medium to large power distance and medium to strong uncertainty avoidance. According to Hofstede's (2001) cultural dimensions, France and the former Yugoslavia are characterised as in the high power distance values whereas the majority of the Western European countries are low on power distance. The cultural value differences between Western and Eastern societies are observable based on the literature review. These differences have been illustrated in relation to their various religious philosophies, their way of life, their interpersonal relationships, their communication type and international patterns.

Cultural differences between Western and Eastern societies have been exemplified through a variety of religious philosophies found in different parts of the world. Low power distance occurs in countries where Anglo-Saxon and Protestant values have significantly affected the culture at large. High power distance occurs in countries that have been influenced more strongly by Catholicism, Hinduism, Buddhism, Islam or Tribalism. Anglo-Saxon Protestantism is generally more egalitarian than other worldviews. Christian religious philosophy and the Ten Commandments teach people to worship God, to love, respect and care for others and to dictate non-violence. Eastern values manifest themselves through religions such as Hinduism and Buddhism (Smart, 1968). For example Buddhism emphasizes common coexistence. Vietnamese Buddhists have been described as people who respect life and practise proper conversation, who control their morality, their own feelings and thoughts, and who do nothing to hurt others. Vietnamese culture is predominantly influenced by Confucianism, which emphasises the value of education, a desire for accomplishment and an obligation to family. Confucianism advocates a common set of presumptions and values. It encourages people to work hard, be responsible, knowledgeable and help others but places a lower emphasis on personal advancement (Jones, 1993). Whilst Shintoism dictated the worship of ancestors, Taoism dictates avoiding social obligations leading to a simple life that is close to nature. It discouraged assertiveness and self-expression, stresses emotional calm and being in harmony with nature.

Cultural differences between Western and Eastern societies have been demonstrated though their approach towards interpersonal relationships. Harmony in interpersonal relations is considered to be an exceptionally important value in Eastern cultures. In these societies, people emphasise self-restraint, avoidance of conflict in interpersonal relations, complaints, negative opinions, criticism and negative emotions. Eastern cultures try to maintain harmony in interpersonal relations, to avoid embarrassment and to "save face" (Dodd, 1987; Huyton, 1991). However, the interpersonal relationships in Western societies are seen as creating friction because people value self-esteem and are less concerned with apologies. For the Vietnamese, it is better to perform acts of self-effacement rather than break group harmony (Leung, 1988). Criticizing in public makes people lose face and damages their relationship with those who criticize. Leung (1991) has indicated that in China one confirms a negative statement with a "yes." In Western cultures, the reaction to failure is often an attempt to improve performance, whereas in Eastern culture, admission to failure and criticism are limited in order to protect the individuals from loss of face. Philippinos avoid disagreement, embarrassment and social disruption that could bring shame (Lynch, 1970). These concepts can appear to be in contrast with the Western emphasis on truthfulness or forthrightness. People in Eastern cultures often apologize in order to maintain social harmony. Having a humble attitude and apologetic nature is considered a sign of good behaviour in Asian culture. The Japanese offer compensation for the other person to maintain harmonious relations while Americans give explanations and justify their acts (Bamlund and Yoshioka, 1990).

Eastern societies are indirect in behaviour, strive to maintain harmony in human relations and follow the ethic of not questioning, disagreeing or hurting the feelings of others. They do not compliment as compliments can cause harm (Leung and Bond, 1984, 1989). Many Westerners are open in their relationships with others and they speak freely about their feelings and personal experiences. For example, cultural differences in managing conflict styles can make collaboration between Westerners and Asian very taxing and demoralizing. For instance Chinese managers may become angry that their Western partners arrange public meetings to make hiring decisions. In addition to irritations that arise when their partners want to hire more overpaid Westerners, they are angry that Westerners want to force their decisions into a public arena rather than in quiet, give-and-take discussions. The Westerners in turn are upset not only that the Chinese exaggerate their ability to operate without them, but with their perceived refusal to deal with the issue. In these complicated conflicts, partners can easily feel stymied and lose confidence that they can work together (Tjosvold, Hui and Law, 2000).

Cultural differences between Western and Eastern societies have also been illustrated in their ways of life. People of the East subscribe to values which are not conducive to economic progress, values that place the family and the community above individual interests, and spirituality above material well-being. By contrast, the West subscribes to the kind of individualism and materialism that gives its people a competitive edge. The Chinese way of life is influenced by the thinking and behaviour of their religious philosophies. For example, Confucianism restricts the expression of emotions. It dictates respect and the ability to foresee how people's behaviour can affect others, the use of precise words, correct naming and speaking with a proper degree of hierarchy (Ryan, 1985). Differences between Western and Eastern approaches are noted in the concept of self (Chung, 1969; Hsu, 1971a, 1971b). Eastern cultures stress belongingness and ego that avoids creating conflict. Chinese are presented as self-sufficient and progress-oriented. The Japanese are presented as independent and open to experience (Jones and Bock, 1960) or as people-oriented, society-oriented, or achievement-oriented, less materialistic and receptive to nature (Stoetzel, 1955; Kikuchi and Gordon, 1966, 1970).

Whilst Western societies are more direct in their behaviour, Eastern societies are very indirect. According to Lustig and Koester (1993), Europeans and Americans emphasize directness and openness in interpersonal interactions and communication. In indirect cultures there is an emphasis on indirectness, ambiguity and the use of third parties and intermediaries (e.g. Japan, Korea, Thailand, Vietnam, China and Africa). The cultural differences between Western and Eastern societies also lead to a different understanding of what constitutes appropriate behaviour and friendship. For instance, qualities such as being "yourself," open, friendly, outspoken, informal, truthful in interpersonal relations that are admired in most Western cultures, are less likely to be admired in Eastern societies as they view Westerners as lacking grace, manners and cleverness (Craig, 1979). People in Eastern cultures are embarrassed by discussing personal matters in relationships. "Opening" themselves to others is an indication of weakness, and that person cannot be trusted. Western friendliness and informality is expressed by referring to others by their first name. However, such informality is regarded as impolite in Eastern societies. Members of Western societies are "touchy" and display their emotions in public. Members of Eastern societies are reserved, behave modestly, and regard Western behaviour as aggressive and unacceptable. For instance, the Chinese do not like to be touched, slapped on the back or even to shake hands. They avoid open displays of affection. Their speaking distance is greater than in the West and they do not appreciate loud behaviour (Wei et al., 1989). What Western societies regard as normal and acceptable behaviour, Eastern societies may regard as insulting and irritating. Eastern people often seem to laugh out of context in the Western view. For instance, the Japanese smile often means embarrassment, instead of happiness. The understanding of friendship is also different. Americans regard friendships as superficial and without obligation. Chinese understand friendships in terms of mutual obligations and reciprocation (Wei et al., 1989).

Cultures have been distinguished on the basis of vocabulary and language used between individuals. The same words can have different meanings and implications. For instance, Eastern societies avoid direct negative answers. "Yes" may mean "no" or "maybe." Western societies mistake Easterners use of "yes" as being untrustworthy. Fontaine (1983) listed difficulties in culturally and linguistically different countries due to differences in word use, styles of interaction and rules governing these interactions. The misinterpretations of words can lead to serious misunderstanding between Eastern and Western societies, particularly when one party tries to be as hospitable as possible toward guests. Language determines what is said and how it is said. This can be observed in the degree of expressiveness and the suitable subjects of discussion with regard to family, religion or politics. Hall and Hall (1987) reported that in high-context cultures, being direct and open, using many verbal codes, expressing emotions or showing feelings may be considered a sign of immaturity. Silence and reserved reactions are necessary to maintain social harmony and not to threaten the face or self-esteem of others. In low-context cultures it is acceptable to express anger, show excitement, and be noisy and confident.

There are many non-empirical as well as empirical studies of the cross-cultural differences between Western and Eastern societies. In general, the most important Eastern values are a strong sense of belonging to groups and society, closeness to the outside world, solidarity, supporting welfare of others, harmonious interpersonal relations, cooperation, sharing, reciprocity, dependency, submission, conformity, loyalty to superiors and ancestors, saving face, suppression of open conflict and competition, living with harmony with nature, belief in supernatural forces, present and past orientation, non-materialistic orientation, in situational behaviour, differences between real intentions and actual behaviour, formal and informal behaviour. Eastern societies are emotionally restrained and sensitive to outside opinions; they value morality, reserve, and cleverness. For them, egoism and selfishness are considered as the worst characters. However, these values are in contrast with the rights of the individual in Western societies. Westerners value individualism, as they believe that each individual is special, unique, and completely different from all other individuals, thus the interests of the individual are paramount. They place importance on learning to be an individual, independent, self-motivated and achievement oriented. They place a high value on personal freedom, material well-being, humanitarianism and equality of people, directness, work, time, success, efficiency and practicality. Since members of Western and Eastern societies have opposite cultural orientations and expectations, they may have a different understanding of what constitutes appropriate behaviour. The cultural differences between members of these societies can directly affect their social interaction with hosts whilst on holiday.

RELATIONSHIP BETWEEN CULTURE, RULES OF BEHAVIOUR AND PERCEPTIONS

Relationship between Culture and Perceptions

Perceptions are based on physiology (the five senses) and also have characteristics related to demography, behaviour, society, culture, economics and psychology (Usunier, 2000). Culture is particularly important as a determinant of perceptions (Samovar & Porter, 1991). It has a great influence on how experiences are perceived and also on the interpreted meaning. McCracken (1986) has referred to "culture" as a lens through which people view the world. Cultural influences may be viewed as how people perceive and assimilate phenomena. Wei et al. (1989) observed that cultural differences lead to different perceptions of what constitutes appropriate behaviour (p. 329). Since views about the world differ, it is unsurprising that perceptions also differ (Robertson, 1970). Perceptions rely on cultural values, expectations, experiences and interests and are culturally determined. Referring to culture as perception, McCort & Malhotra (1993) noted that "culture is the shared, consumption relevant knowledge system necessary to operate in a manner acceptable to one's society. This knowledge system, though the formation of culturally learned rites of perception and interpretation imbues objects and behaviours with meanings for its members." Urriola (1989, p.66) indicated that culture is "the sum of peoples' perceptions of themselves and of the world. ..." Triandis (1972) defined the main elements of subjective culture are values, role perceptions, attitudes, stereotypes, beliefs, categorizations, evaluations, expectations, memories, and opinions. He mentioned that the similarities in subjective culture lead to frequent interaction among members of similar cultural groups. He also reported that "when the similar behaviour patterns obtained in one culture differ from the similar patterns obtained in another, we infer the existence of some differences in subjective culture" (Triandis, 1972, p.9). This definition was confirmed by the work of Samovar et al., (1981) who noted that the members of a similar subjective culture have similar values, conform to similar rules and norms, develop similar perceptions, attitudes and stereotypes, use common language, or participate in similar activities.

Culture affects each stage of the process of perception. Initially, it provides patterned material for perception (e.g. architecture, the aroma of foods, and the sound of music). Later, through verbal and nonverbal means, it suggests the proper labelling of and responses to perceptions of patterns. The relationship between culture and perceptions has been frequently noted. Several empirical studies have identified the influence of culture on

Literature Review

perceptions (Mayo & Jarvis, 1981; Schneider & Jordan, 1981). Redding (1980) and Mayo & Jarvis (1981) pointed out that culture causes different nationalities to perceive differently with those growing up in different environments perceiving differently because they interpret causes differently (Segall et al., 1990). Richardson & Crompton (1988) attributed the different perception of French and English Canadians to cultural differences which elicit different responses to market strategies. Singer (1982) pointed out that different cultural values lead to different perceptions. One example is aesthetic values which are culturally determined and influence the perception of physical appearance and attractiveness. Keown et al. (1984) reported cultural influences on the differences between tourist perceptions of retail stores in 12 countries. Huang et al., (1996) found significant differences in individual perceptions of leisure activities prompted by different personal values and cultural backgrounds. People with significantly diverging personal values exhibited significantly different perceptions.

Tourist perceptions have become a focus for researchers who are involved in examining the various dimensions of the tourist perspective. The findings of previous cross-cultural research has confirmed that tourist perceptions of a destination or hospitality businesses may vary on the basis of country of origin (Richardson & Crompton, 1988; Catalone et al., 1989; Huang et al., 1996). Pizam & Sussmann (1995) investigated tour guide perceptions of similarities and differences between tourists from four countries. The same survey was subsequently repeated among Israeli tour guides (Pizam & Reichel, 1996). In both studies, tour guides perceived that different behavioral characteristics were evident amongst tourists from different nationalities.

A number of studies have examined host perceptions of tourists. Brewer's (1984) study of Mexico concluded that local residents have "general" stereotypes of all Americans, which lead to "specific" stereotypes which are then applied to American tourists. Pi-Sunyer (1978) found that Catalans stereotype English tourists as stiff, socially conscious, honest, and dependable. Boissevain & Inglott (1979) observed that the Maltese characterized Swedish tourists as misers, and the French and Italians as excessively demanding. Other studies found that residents of tourist destinations perceived tourists to be different than themselves in a variety of behavioural characteristics and lifestyles. Pizam & Telisman-Kosuta (1989) found that in the destinations where a majority of tourists were foreigners, the residents perceived the tourists to be different from themselves in a variety of behavioural characteristics, such as attitudes or morality. In destinations where the majority were domestic tourists, the differences between the tourists and the residents were perceived as minimal. Similarly, Wagner (1977) in a study of charter tourism to the Gambia noted that the locals saw Scandinavian tourists as a "clearly demarcated group, whose dress, behaviour and life-style set them apart" (p. 43).

Culture has also been empirically proven to have an impact on the formation of expectations (Armstrong et al. 1997) as well as quality expectations (Luk et al., 1993). The cultural differences in expectations regarding service levels between hosts and visitors left many with negative impressions" (1989, p. 3). In the same vein, (Befu, 1971) reported that Japanese hosts take good care of the affairs of their guests in advance, anticipate the guests' needs and believe that they know best what the guests' needs are. However such an attitude may also be frustrating for Western tourists who think they know best what their needs are. Western tourists may regard Japanese hospitality as uncomfortable. On the other hand, the Western tradition of not anticipating the guests' needs in advance may negatively impact on the satisfaction of Japanese tourists with the hospitality of the Western hosts.

Additionally, the relationship between culture, perception and interaction was also highlighted in a number of studies. Sheldon and Fox (1988) reported that poor quality service may lead to unpleasant encounters between tourists and hosts, low morale, and unfriendly attitudes. They indicated that "interacting with service personnel is a primary way in which visitors form an impression and make judgments about their hosts. In Reisinger & Turner's (2003) study, cultural value is seen to impact on interaction behaviour which is important in services given the nature of the service encounter, which is a dyadic interaction between Asian tourists and Australian service provider.

There appear to be a number of factors creating serious problems for how tourists and host perceive each other. It is important to determine how tourist perceptions of hosts and vice versa particularly in the cross-cultural context of the tourism industry. Cultural differences mean that there is a considerable negative perceptions arising amongst Western and Eastern tourists, as well amongst Western tourists of Eastern hosts and vice versa. Insensitivity to these differences may cause misunderstanding and interaction difficulties between tourists and hosts. Although there are many other cultural differences between members of Western and Eastern societies that have impact on their perceptions and social interaction, it can readily be seen that members of Western and Eastern cultures have totally opposite cultural orientations and expectations from the perception and social interaction. These cultural differences may have a negative influence on how people perceive other.

There is a significant relationship between culture and perceptions since perceptions of the world around are influenced by the culture into which one has been socialized. It is important to understand cultural value orientations that affect perceptions. Most tourism and hospitality industry employees appear to implicitly or explicitly acknowledge the existence of tourist cultural differences in terms of interests, needs, expectations, destination or hotel selection and preferred activities. Clearly destination image and perceptions influence vacation choice decisions and national cultural characteristics affect tourist perceptions.

Intercultural Interactions

The relationship between culture and rules of interactions has been highlighted in previous studies. Sheldon and Fox (1988) have noted the role of cultural differences in the patterns of interaction between guests and service providers. Such differences may lead to diverging perceptions about what constitutes proper treatment of guests and shape the attitude of hosts towards tourists (Richter, 1983). Wei et al. (1989) emphasized the influence of cultural differences on the interactions between service providers and visitors. Poor quality service may lead to unpleasant tourists/hosts encounters, to low morale and to unfriendliness. Interacting with service personnel is important for visitors in forming impressions and making judgments about their hosts. Reisinger and Turner (1998, 2002a & 2002b) found that cultural values impact on interactions between Asian tourists and Australian hosts. This is an important finding because service involves dyadic interactions between customers and service providers. File et al. (1992) have argued that service encounters are affected by cultural perceptions and values, thus strengthening the need to consider customer participation in light of cultural values.

Products and services offered to tourists are distinct from those provided in other service industries and require distinct managerial approaches (Lovelock, 1991). Tourism extends beyond physical resources and "tangible" offerings (such as hotel lobbies, bed and furnishings), to the conviviality of service providers. Customer perceptions of the service encounter are critical (Sparks and Callan, 1992; Sparks, 1994).

Service quality is a "phenomenon which contributes to the strength of interpersonal, intra-organizational and inter-organizational service encounters" (Svensson, 2001, p. 357). Marketing professionals view service quality as critical to the development of fruitful relationships. In the context of a dyadic service encounter the service quality is dependent on an interactional process between customer and service providers. This may be described as "a theatre, a show, or an act of performance" (Svensson, 2001, p. 357). Svensson has argued that the quality and experience of service depends on the performance of both actors. He suggests that the "better the interactive process, the better the outcome that may be achieved in the service encounter" (Svensson, 2001, p. 368). A tourist who interacts with a flight attendant, a tour guide or a waiter becomes a co-producer. Productivity and service quality are dependent on contributions from both the customer and the service providers. Interactions may be intense and intimate.

Service quality is connected with subjective perceptions of service by customers. Service encounters occur between customers and service providers. Direct person-to-person tourist interactions are frequently intense. Interactions depend on the interpersonal skills of staff and the environment within which such services are delivered. The cultural values and rules of behaviour of tourists and hosts determine their satisfaction and the quality of their interactions (Pearce and Moscardo, 1984). Service providers, marketers and researchers should examine the "impact of cultural differences on the quality of cross-cultural interactions between tourists and locals working in the tourism industry" (Dimanche, 1994, p. 127). An examination of this issue can make an important contribution to overall tourist satisfaction.

Service encounters involve interactions between customers and service providers. According to Goffman's (1961) role theory, interactions during a service encounter are principally determined by the respective roles of the customer and the provider. Service encounters involve role performance, namely human interactions with clearly defined short-term goals that are agreed upon by society. Haring and Mattsson (1999) have noted that linguistic communication is the primary instrument of human-to-human coordination in social activities. They maintain that service providers must understand communication behaviours that occur during the service encounter and which impact on perceptions of service quality. Haring and Mattsson (1999) view quality as being influenced by the relationship between those involved in the communication and social setting. Effective communication involves "the totality of contextual characteristics of an activity that bears on its participants' ability to achieve the activities stated or implied purpose" (p. 32). Service providers who set out to manage behaviour and create a perception of good quality should understand the contextual influence of the encounter on communication behaviour.

Haring and Mattsson (1999) identified two major conversation types which occur during service encounters: professional or task related and social or non-task related. Professional interactions are about accomplishing tasks, are well-defined and are stable over time. Social interactions do not involve a clear-cut purpose and may contribute to building a perception of trust and security. The two types of interactions can overlap and depend on context. The authors describe social activities as "the main causal factor for co-coordinating human interaction" (Haring and Mattsson, 1999, p. 33). In examining service encounters between hosts and guests, the present study considers both professional and social interactions.

The foregoing review has shown that when members of two or more cultures communicate, different values and rules of social interactions increase the likelihood of misunderstandings. Since social or professional interactions are based on the culturally conditioned perceptions of others, people bring their unique perceptions of the world to the social or professional contacts that they have with others. Interactions are informed by culture, which shapes how individuals perceive others. Cultural variations help to explain perceptual differences. Cultural differences can lead to problems in cross-cultural social or professional interactions. Tourism interactions may be viewed as 'intercultural', depending on the degree of heterogeneity between the cultural backgrounds of those involved on their patterns of beliefs, rules of behaviour, verbal and non-verbal behaviour, perceptions, and attitudes. Those belonging to a common culture are likely to have more features in common than those who belong to different cultures. Members of similar cultural groups commonly speak the same language, share the same religion, experiences, perceptions, and have similar world views. According to Pearce and Moscardo (1984), quality of satisfaction and interactions is determined primary by closeness of the cultural values systems of the tourist and host.

Rules of behaviour and interpretations of their appropriateness in social situations differ in various cultures. These rules can be used as a starting point for analysing differences between cross-cultural interactions. Service interactions within international tourism involve direct face-to-face encounters between people from different cultural backgrounds. Social or professional interactions between tourists and hosts in the context of a service encounter play an important role in shaping tourist perceptions towards service attributes and performance. Satisfaction with such interactions depends on service quality and contributes to overall holiday satisfaction.

The influence of rules of behaviour on host-guest service interactions has received minimal attention from researchers. No previous study has examined cultural interactions and the between hosts and guests in tourism service settings in Vietnam. In the current study, interactions involve French tourists and Vietnamese service providers such as receptionists, waiters, tour guides, bus drivers or shop assistants.

SERVICE QUALITY AND SATISFACTION

Measuring and Evaluating Service Quality

Service quality is an elusive, abstract and intangible construct that is difficult to define and measure (Parasuraman et al, 1991). Identifying the relationships between service quality, customer satisfaction and purchasing behaviour remains largely speculative. Several leading instruments for measuring service quality are still viewed as controversial including SERVQUAL, SERVPERF and Non-Difference Score. However for the purposes of the current study, SERVQUAL has been chosen.

The word 'SERVQUAL' is an abbreviated form of service quality. The measurement was developed by a leading group of marketing researchers, Parasuraman, Zeithaml and Berry in 1985. The basic theory of the measurement was Oliver's (1980) research expectation-performance confirmation model. Interestingly, different understandings about customer expectation between Parasuraman et al. and Oliver have been acknowledged. As mentioned in the above section on expectation. Oliver (1980) defined expectations as consumer-defined probabilities of the occurrence of positive or negative events if the consumer engages in some behaviour. However, Parasuraman et al. understood the term expectation as the customer desired level of performance, for instance where the customer engages in some behaviour (Lee and Kim, 1999).

SERVQUAL was originally composed of ten components which were subsequently collapsed into five broader dimensions: Reliability, Tangibles, Responsiveness, Assurance, and Empathy. These had in turn been derived from earlier exploratory work done by Parasuraman et al. in 1985 which aimed to uncover the dimensions that consumers used to formulate expectations about and perceptions of services. In subsequent, SERVQUAL analyses focussing on four service industries namely banking, credit card services, repair and maintenance and long distance telephone.

- *Reliability:* Ability to perform the promised service dependably and accurately
- *Tangibles:* Physical facilities, equipment, and appearance of personnel
- *Responsiveness:* Willingness to help customers and provide prompt service
- *Assurance*: Knowledge and courtesy of employees and their ability to inspire trust and confidence
- *Empathy:* Caring, individualised attention the firm provides its customers

SERVQUAL was developed by Parasuraman et al. (1988) based on these dimensions of service quality. Each of these dimensions was made up of four to five items. The SERVQUAL instrument is made up a total of 22 items to measure expectations and perceptions of service. It is administered twice in different forms, one to measure expectations and one to measure perceptions. The respondent service quality score is then derived by subtracting the perception score from the expectation score as follows:

Service Quality (SQ) = Perceptions (P) - Expectations (E)

Parasuraman et al. (1988) have claimed that the five dimensions of service quality are generic. In their later work on SERVQUAL, Parasuraman et al. (1991) modified the wording of the expectation items. Whilst the 1988 version of SERVQUAL aimed to capture respondent normative expectations, subsequent versions focused on what customers *would* expect from excellent service companies. The authors argued that measuring customer normative expectations by using 'should', might contribute to unrealistically high expectations scores (Parasuraman et al., 1991).

SERVQUAL is the most generally accepted representation of the service quality construct. Following Parasuraman et al.'s pioneering work,

SERVQUAL has been widely replicated in various service industries including utilities (Babakus and Boiler, 1992), dental services (Caiman, 1990), apparel retailing (Gagliano and Hathcote, 1994) and business school placement centres (Caiman, 1990).

Though Buttle (1996) has noted that SERVQUAL has been widely applied and is highly valued, there have been some criticisms of the SERVQUAL instrument. Critics have suggested that despite its popularity, its usefulness very limited (Carman, 1990; Lee and Kirn, 1999). Brown et al. (1993) reported that "SERVQUAL needs to be customized to the service in question, in spite of the fact it was originally designed to provide a generic measure that could be applied to any service" (p. 128). They suggested additional items and amendments to the wording. Confusion was noted about the precise meaning of expectations. Parasuraman et al. (1991) define expectations as including the following: desires, wants, what a service provider should offer, normative expectations, ideal standards, what the customer hopes to receive and adequate service. The multiple definitions and corresponding measurement operationalisations of the SERVQUAL model have led to an expectation norm concept that is loosely defined and open to multiple interpretations.

Association between Service Quality and Satisfaction

The various definitions of satisfaction and performance quality are broadly consistent with the expectancy-discontinuation paradigm. This paradigm is derived from two processes: development of expectations about service outcomes, and the disconfirmation judgment that results from comparison of the outcome against these expectations. Confirmation arises when actual performance matches initial expectations. When the actual performance exceeds or falls short of expectations, positive or negative disconfirmation results. Positive discontinuation leads to satisfaction, while negative disconfirmation leads to dissatisfaction. Tourist satisfaction is generally derived from the expectancy-disconfirmation paradigm and from a comparison of expectations and actual performance.

Since service quality and satisfaction draw upon the expectancy disconfirmation paradigm, there is a degree of confusion about how to distinguish between the two constructs. Whilst it is generally agreed that service quality and visitor satisfaction should be treated differently, no fundamental distinctions have been established between the theoretical derivations of the two constructs. As a result, there are various conceptual interpretations of service quality and satisfaction, and of the relationship between them in the marketing, recreation and tourism literatures. Different conceptualizations of service quality and satisfaction are discussed in the following section.

Service Quality and Satisfaction as Synonyms

According to Patterson (2000), Patterson and Johnson (1993) and Patterson, Johnson and Spreng (1997), some researchers have used the constructs of service quality and satisfaction interchangeably in the belief that differences between them are semantic rather than substantive. The lack of precision in the literature may have prompted some researchers to treat service quality and visitor satisfaction as a single construct.

Ohio State Parks (1996) developed a questionnaire intended to investigate visitor satisfaction with services of the state park. The scale used in the questionnaire to measure satisfaction, measures the quality of the service offered. The quality of the park management's performance was measured including any service weaknesses. Based on the results, the researchers mistakenly concluded that the questionnaire provided feedback on "levels of customer satisfaction with their outdoor recreation experience in Ohio State Parks" (p. 2). In other cases, service quality is used as a measure of satisfaction. Howat et al.'s (1996) study conducted in Australia, evaluated tourist satisfaction by measuring the effectiveness of sport and leisure center management, using indicators of customer service quality (CSQ). In the questionnaire, the five dimensions of service quality identified by Parasuraman et al. (1985) were equated with "dimensions of customer satisfaction," and served as "points of reference from which the performance indicators effectiveness (CSQ) questionnaire was developed" (p. 83).

Service Quality and Satisfaction as Distinct Constructs

A degree of consensus has recently emerged among researchers that service quality and visitor satisfaction should be viewed differently, although there is no clear consensus about the critical differentiating elements. Parasuraman et al. (1988) argued that although both constructs involve a comparison between expected and perceived service, the expectation standards in the two constructs are different. They insisted that the predicted service is the comparison standard in customer satisfaction (i.e., expectation of what the service would be) while the standard in service quality is the ideal or desire (i.e., what the service should be).

Conceptualizing service quality and visitor satisfaction as synonymous is illogical. Within this approach, it is implied that whenever visitors have a perception of high quality of service, they will have high levels of satisfaction and vice versa. This is often not the case in practice. In some cases, a high satisfaction outcome may arise even when perceived service quality is low. Low levels of satisfaction may arise when the perceived service quality is high. Tourists may have recently had a bad experience while travelling to a site (i.e., being booked for a traffic infringement). In such circumstances, they will not be in the right mood to be satisfied, even if they were to receive high quality service. Tourist satisfaction and service quality are not necessarily positively correlated. Tourist satisfaction may be low (high) when service quality is high (low). It may be concluded that tourist satisfaction and service quality are two different constructs.

SERVICE QUALITY: HOST ATTRIBUTES AND PERFORMANCE

Service is a component of product delivery and may be defined as "any activity or benefit one party can offer to another that is essentially intangible and does not result in the ownership of anything. Production may or may not be tied to a physical product" (Kotler et al., 1989, p. 725). According to Parasuraman et al. (1990), the purchasers of services such as tourism may experience greater difficulty in evaluating quality than is the case for the purchasers of tangible products. This is because services have three main unique features.

The first feature is *intangibility* on the basis that services deliver performances and experiences rather than the objects. Services cannot be seen or tried prior to purchase but only during or after consumption. Service purchasers are generally likely to perceive more risk than the purchasers of goods.

A second feature is *heterogeneity* which acknowledges that service delivery may be inconsistent across individuals, time and situations. The production of goods is usually based on standardized criteria in order to achieve uniformity, while services can rarely be standardized. A travel agency for example, may set rules about customer service, but the services delivered by individual employees may differ. In this case, consumers do not possess standardized criteria to evaluate the service that is performed.

A third feature is *inseparability*. Unlike the production of goods, the purchaser is usually involved in the production process and quality is often determined on the basis of service delivery. A service is consumed while it is produced and visitors are not in a position to evaluate service quality before the service is delivered. Quality occurs during or after consumption.

According to Parasuraman et al. (1988), service quality is the result of a subjective customer perception of service. Service quality is often related to the customer perception of a service. Service perception often refers to the perception of the interaction between a customer and a service provider. Grönroos (1982, 1990) reported that service quality is made up of two components, namely technical quality and functional quality. Technical quality refers to the performance received by visitors, for instance a cash withdrawal at a bank. Functional quality refers to the process of service delivery. The withdrawal may be provided through an ATM or in person by a teller. Similarly, Lehtinen and Lehtinen (1982) proposed three dimensions of quality including physical, corporate and interactive quality. Physical quality relates to the technical aspects of the service and the latter two dimensions emphasize the corporate image of the service organization and the interactive processes occurring between a tourist business and its visitors. As service quality occurs during the service encounter between a consumer and a service provider, the interactive dimension of service quality is central and the quality of such interactions is essential for the provision of quality service (Parasuraman et al, 1985; Solomon et al., 1985; Urry, 1991). Martin (1987) distinguished between the procedural and convivial dimensions of service quality. The former deals with systems of selling and distributing a product to a customer and is mechanistic in nature. The latter highlights service provider behaviour, courtesy, attentiveness, friendliness, their verbal and non-verbal skills and positive attitudes or personal interest towards their customers such as being appreciative of the customer or fulfilling the customer's psychological needs. According to Martin (1987), this convivial dimension emphasises the customer's need to be respected, and to feel relaxed, comfortable, important, pampered, and welcomed.

Evaluations of service quality embrace not only the service delivered, but also the manner in which it is delivered. In the tourism sector, perceptions of service quality rely heavily on positive perceptions towards hosts (service providers), as perceptions of hosts are part of the overall perceptions of a tourism product. Host attributes are the fundamental aspects of service quality. For instance, providing prompt and courteous service to clients or smiling in a pleasant and involved way to customers are the important attributes of service quality. Crompton and MacKay (1989) defined service quality as the quality of service attributes. Service attributes are the constituent elements of the opportunities that management provides for tourists. These are controlled and manipulated by tourist suppliers. To reflect this perspective, Crompton and Love (1995) used the expression "quality of opportunity" to describe service quality in the tourism field. Quality of opportunity or performance quality of the tourism entity is defined as the quality of services that are under the control of service suppliers. It is operationalized as the disparity between the desired level of service and perceptions of the performed level of service. Pizam et al. (1978) highlighted employee friendliness and courtesy towards tourists as well as their willingness to help as the important hospitality characteristics in the service delivery process. Callan (1997) defined service quality as "staff who get things done promptly and provide honest answers to problems"; as "a responsive, caring and attentive staff" who "making the recipient feel thoughtful, efficient, correct and magnanimous"; or as "a hospitality which leads the guest to feel at home" (p. 48). Saleh and Ryan (1992) also emphasised that "appearance is not only important but to some extent is more important than the range of facilities being provided" (p. 168).

Pearce (1982a, 1982b) demonstrated that overall tourist perceptions of service will be determined by interactions with a variety of people within the tourism industry. These include hoteliers, restaurateurs or other employees who contribute to the overall tourist perceptions of service. Sutton (1967) reported that competency in providing services is an important element influencing positive tourist perceptions of service. Negative tourist perceptions arise because of impoliteness on the part of service providers, their feeling of discontent or because of not being able to achieve a certain standard of service. Therefore, the positive service perceptions such as the provider friendliness or politeness encourage both repeat consumption and repeat visitation to the host region. By way of contrast, negative service perceptions, which are created by variables such as impoliteness or annoyance at poor service, will produce the opposite effect.

Service quality is multidimensional and difficult to evaluate. With their subjective characteristics, each service quality dimension may be perceived differently depending upon whether it is perceived by a visitor or by a service provider particularly in a cross-cultural context. According to Gee (1986), tourist perceptions of hosts are the most important of the various tourist perceptions. Studies conducted for tourist destinations have investigated customer service quality evaluations, but little cross-national exploration has been undertaken on the influence of cultural values and rules of behaviours on tourist assessments of service attributes and performance. No previous study has investigated tourist cross-cultural perceptions in term of service quality in Vietnam. For the purposes of the current investigation, tourist perceptions of hosts will be assessed in terms of service providers' attributes and performance. An examination of tourist perceptions of host attributes and performance should enable the detection of better negative perceptions, change or modify them if necessary, and therefore, respond better to the diverse culturally needs of tourists. Assessing the perceptions of French travellers towards Vietnamese hosts should be valuable, enabling the salient attributes and the re-evaluated image to be incorporated into tourism marketing.

Host-Guest Mutal Perceptions

According to Argyle (1990), perceptions of people are more complex than perceptions of other physical objects, since the sensory inputs are normally obtained as part of the process of interaction. People perceptions have been studied extensively in recent years including by Smith (1967) and Tagiuri (1969). However, much of the research has viewed perceptions of people as being essentially a cognitive issue, rather akin to concept-formation or problem- solving. In the context of the current study, the term 'perceptions' is used differently from its use in experimental psychology. It is not for example concerned with whether guests perceive hosts or vice versa to be dark or fair skinned. It is more concerned with host inferences about guests and with guest inference about host personalities, abilities and attitudes during the service encounter: namely with inferences about another person based on visible or audible behaviour. Inferences are made because they are needed by each of the interacting parties. The nature of these inferences will depend on the situation and on the relationship between those who are interacting (Argyle, 1967). Hargie (1986) indicated that perceptions are important in social relations and human interactions and are "the impressions that people form of one another and how interpretations are made concerning the behaviour of others" (p. 47). He reported three types of perceptions which influence social interactions: (1) perceptions of other people (tourist perceptions of hosts and host perceptions of tourists); (2) meta-perceptions (or perception of the perceptions); and 3) perceptions of one's own (host perceptions of themselves and tourist perceptions of themselves). Perceptions of other people (inter-perceptions of Vietnamese hosts and international guests towards service quality) were chosen in the current study. People perceptions of each other can be positive or negative, depending on the environmental influences of judgment such as culture. Robinson and Nemetz (1988) pointed out that cultural similarities bring people together and cultural dissimilarities separate them. They reported that negative perceptions occur amongst people with cultural differences because of the degree of cultural similarity or dissimilarity in signals and meanings and the knowledge of the other culture and its influence on meanings of the signals. For those reasons, negative perceptions existing between the perceivers are high when the cultural dissimilarities between them are big. The perceived cultural dissimilarity in values therefore produces perceptual mismatching in interpretations and consequently creates cross-cultural misunderstanding between people.

Perceptions are derived from consumer behaviour research and carry the greatest weight in various decisions made by tourists including destination choice, consumption of commodities while on vacation, satisfaction and intention to revisit. Perceptions are the subjective reality for consumers. Tourist perceptions of hosts are important because they influence both decision-making behaviour and destination selection. Host perceptions of tourists are also essential because they are the key point for identification of tourist perceptions of hosts. Hosts may be conceptualised in different ways. Several studies have referred to hosts as service providers, local residents or people employed in the tourism trade. For the purposes of the current study, hosts are referred to Vietnamese service providers and the terms will be used interchangeably. Host-guest mutual-perceptions of service attributes and performances, and the host-guest social/professional interaction are examined in this current study.

Tourist Perceptions of Hosts

Perceptions of hosts form part of the overall perception of tourism products. The tourism industry relies heavily on the formation of positive perceptions towards hosts. As a customer often relates service quality to a perception of service, satisfaction with service quality thus can be measured by tourist perceptions of hosts. This is because the attributes and performance of service providers are vital elements in service delivery. As tourists are in close contact with hosts, their perceptions towards service provider attributes and performance form an important part of overall destination perceptions. Positive perceptions of tourists towards hosts can influence their destination selection. Positive perceptions may contribute to overall perceived holiday experience, affect overall holiday satisfaction, and in turn motivate future behavioural intentions.

Perceptions have become a common focus of study amongst researchers examining looking at various aspects of the tourist perspective. The findings of past cross-cultural research have confirmed that tourist perceptions of a destination or hospitality business may vary on the basis of countries of origin (Huang and Wu, 1996; Armstrong et al., 1997). Anastasopoulos (1992) surveyed Greek tourists visiting Turkey and measured how their attitudes towards Turkish people and institutions changed by comparing their beliefs before and after their visits. His findings showed a negative direction on tourism's role as a mediator of change in attitudes towards other nationalities, more specifically, visiting a country for tourism purposes can have a negative impact upon one's perceptions of the host nation. De Albuquerque and McElroy (2001) investigated how foreign tourists perceive the behaviour and attitude of the local people of Barbados. Harlak's (1994) study revealed that Germans, Americans and British were generally perceived with their positive characteristics, but British tourists were perceived to be negative for the category 'miser-generous', differentiating themselves from other nationalities in the study. Harlak hypothesized that "the tendency to perceive tourists as an economic object is higher than the tendency to perceive them as human beings" (1994, p. 148).

Numerous studies have emphasised the importance of tourist perceptions of host communities. Ross (1991) noted that tourist perceptions of local people are an important variable in the decision to re-visit. Pearce (1980, 1998) indicated that the degree of liking or disliking of people encountered influences tourist destination perceptions. Positive or negative perceptions towards hosts influence destination choice. According to Woodside and Lysonski (1989), destinations with strong, positive images are more likely to be considered and chosen during the travel decision processes. Likewise, tourists will enjoy hosts more if they hold positive perceptions towards service providers and local people. And the more favourable the perception of the hosts, the greater the prospect that tourists will choose particular hosts from similar alternative destinations, to experience host products and services. Therefore it is essential for local residents and service providers to develop positive perceptions in the minds of potential tourists. Once they have developed, negative perceptions are durable and hard to change.

In order to avoid negative perceptions of people from different cultural backgrounds, it is essential to understand the target culture, how and why people from these cultures perceive the way they do, the reasons for cultural dissimilarities, and consequently modify one's own cultural understanding (Robinson and Nemetz, 1988). Tajfel (1969) reported that perceptions are influenced by cultural similarity and familiarity. Acceptance is determined on the basis of perceived cultural similarity and familiarity of values (Rokeach et al., 1960) and the liking of others (Freedman et al., 1981). Cultural familiarity with physical appearance generates positive affiliation between people. People desire to associate with those who are physically attractive to them, and will give such people preferential treatment (Huston and Levinger, 1978). Perceived cultural dissimilarity and lack of familiarity results in negative perceptions and inhibits affiliation (Robinson and Nemetz, 1988). Pizam and Telisman-Kosuta (1989) found that in destinations where most of tourists are foreigners, residents perceived the tourists to be different from themselves in a variety of behavioural characteristics such as attitudes or morality. However, in destinations visited mostly by domestic tourists, residents perceived only small differences. As a result, it is essential to know how differences in cultural background determine differences in perceptions, and which cultural factors create positive and negative perceptions. Such understanding could facilitate the development of positive perceptions and deepen the understanding between people from different cultural backgrounds.

Host Perceptions of Tourists

Many studies have investigated tourists' perceptions of their hosts. According to Pearce (1988), host perceptions of tourists have been studied more frequently than tourist perceptions of hosts. Host perceptions of tourists which may be positive or negative are an important point of identification for tourist perceptions of hosts. The English Tourist Board

(1978) noted that Londoners were happy about the tourist presence despite the negative social effects of tourism development. Rothman (1978) noted that American hosts developed friendships with travellers even though they complained about the increased carrying capacity of tourists. Boissevain (1979) reported positive perceptions of tourists and the development of friendships between tourists and hosts in technologically unsophisticated countries.

Negative host perceptions of tourists have also been noted in the literature. Pi-Sunyer (1978, 1982, 1989) reported that mass tourism can result in a growing lack of concern and empathy amongst hosts towards tourists and that the negative perceptions can lead to discrimination towards tourists in the area of services and prices. As tourism expands, facilities provided for the benefit of visitors can be expected to arouse a degree of opposition and discontent amongst residents (Butler, 1980). There is a perception that tourists are responsible for deteriorating standards of living. As tourist numbers increase, hosts feel social stress and became resentful due to a loss of privacy (Smith, 1989). Feelings of anxiety, jealousy, xenophobia, disinterest, rudeness and even physical hostility may develop among host communities (Pearce, 1982b, 1982c). Positive tourist perceptions progressively turn into negative stereotypes. In a study conducted in Mexico, Brewer (1984) suggested that local residents have "general" stereotypes of all Americans, which lead to "specific" stereotypes applicable to American tourists. Pi-Sunyer (1978) found that Catalans stereotyped English tourists as stiff, socially conscious, honest, and dependable. Boissevain and Inglott (1979) observed that the Maltese characterized Swedish tourists as misers, and the French and Italians as excessively demanding. All of these factors create serious problems in the perception of tourists. Consequently, it is essential to determine the host perceptions of tourists and whether or not hosts accept tourists as all these factors could create serious problems in the way hosts perceive tourists.

It is clear that tourist destination images and perceptions influence tourist vacation choice decisions and that national cultural characteristics will affect tourist perceptions. Host perceptions of tourists provide a point of identification for tourist perceptions of hosts. Tourist perceptions of hosts are the most important element of overall tourist perceptions (Gee et al., 1989). An examination of tourist perceptions of host products and services will enable detection of negative perceptions. If necessary, these can then be changed or modified, allowing a better response to the needs of culturally different tourists.

To date, little cross-national exploration has been undertaken on the influence of culture upon host-guest service encounter interactions. Moreover, no study has investigated the cross-cultural mutual perception of hosts and guests in term of service attributes and performance in Vietnam. This book investigates how the cultural values and rules of behavior will impact on host-guest service interactions, and also the importance of service quality as perceived by hosts and guests. Specifically, it investigates tourist perceptions of expectations from services and service encounters will be identified as well as those of the Vietnamese hosts and factors determining these perceptions. Host-guest service encounter interactions is considered as one component of tourist perceptions of host service attributes and performance. The social/ professional interactions between tourists and service providers is an important part of every service encounter and can contribute to tourist overall holiday experience and satisfaction. An examination of tourist perceptions of service provider attributes and performance and host products should enable an improved detection of negative perceptions with a view to changing or modifying them if necessary, and responding better to the diverse cultural needs of tourists. Assessing the various perceptions of French travellers towards the Vietnamese service providers should be valuable, enabling the salient attributes and the re-evaluated image to be incorporated into tourism marketing planning.

Chapter 3

METHODS

METHODOLOGICAL PROBLEMS IN CROSS-CULTURAL RESEARCH

Cross-cultural research involves a comparative approach (Venkatesh, 1995) and has been used in many behavioural studies. The uniqueness of cross-cultural research lies in its comparison in different cultural settings. Cultures shape consumer behaviour while being shaped by consumer attitudes and behaviours (Van Raaij, 1978). One of the characteristics of cross-cultural research is that there is no control over the distribution of the sample. Sometimes this leads to unwanted differences. In most instances, it is not appropriate to apply the same research instrument samples drawn from multiple cultures (Brislin, Lonner and Thorndike, 1973). Another challenging factor of cross-cultural research in consumer behaviour is that it has been developed primarily in Western countries based on western concepts and instruments. This may introduce ethnocentrism in the types of questions, in the concepts employed and in the explanation given to the results (Berry et al., 1992).

There are numerous alternative measurement techniques and analytical procedures available for the examination of values. Many have been criticized partly because it is often difficult to measure and analyse values in cross-cultural research. Values are: 1) abstract constructs and not easily observed; and 2) difficult to translate into different languages as their interpretations depend on the cultural backgrounds of the respondents and researchers.

According to Damen (1987), it is often the case that "values ... and evaluations of the behaviours of strangers are based on the values and belief system of the observers" (p. 192). Often there is confusion between values and other related concepts and of matching researcher value interpretations and respondent behaviours. Feather (1980b) reported on a number of studies which claimed to measure values, but were in practice assessing specific attitudes and interests. The difficulties involved in choosing which values should be assessed add to the difficulties of measuring values (Rokeach, 1973). Findings from previous studies have also suggested values central to individual high rankings. For instance, Atkinson and Murray (1979) found that social values such as love, family, or friendship were given a higher priority than economic values. Bharadway and Wilkening (1977) and Chamberlain (1985) suggested that leisure values have been ranked lowly and recommended focusing on values which are less central to the individual. As a result, the only techniques appropriate particular to cultures are those comparable or equivalent across cultures (Feather, 1975). According to Feather (1986a), studies using multiple methods of measurement are best placed to understand cultural values. The choice of technique to measure values in cross-cultural research also creates problems. These problems also concern the *emic* versus the *etic* approach, the appropriate equivalence of measures and meanings, the ways of maximizing reliable and valid measurements, and the logic of comparative analysis.

The *etic* methodology compares one culture with another. Researchers who follow an *etic* approach in cross-cultural consumer research are generally looking for universal or culture-free theories and concepts. They search for variables and constructs common to and comparable across all cultures in order to discover differences or similarities between cultures. This approach is commonplace in cross-cultural psychology and in other comparative social sciences. An alternative approach is the *emic* approach, which focuses upon understanding issues from the viewpoint of the subjects being studied. Based on this approach, culture is firstly defined emically as the 'lens" through which all phenomena are seen. It determines how these phenomena are apprehended and assimilated. Secondly, culture is the `blueprint' of human activity. It determines "the coordinates of social action and productive activity, specifying the behaviours and objects that issue from both" (McCracken, 1988, p. 73). *Emic* approaches to culture do not intend to compare two or more differing cultures directly, but promote a complete understanding of the culture of study through "thick description" (Geertz, 1973). The methods utilized in conducting *emic* research do not provide "culture-free" measures that can be directly compared. Instead, they provide "culture-rich" information. From an applied perspective, these two definitions of culture *emic* and *etic* can be considered as two sides of the same coin (Luna and Gupta, 2001). Culture is a lens, shaping reality, and a blueprint, specifying a plan of action. At the same time, a culture is unique to a specific group of people. By utilizing the research provided by both approaches, researchers can gain a more complete understanding of the culture(s) of interest. The choice of *emic* versus *etic* approaches depends on several important factors including the purpose of the study, the nature of the research question and the researcher's resources and training. The *emic* approach has been chosen for the purposes of the present study.

Research Methods and Sampling Frame

This study investigates the perceptions and satisfaction levels experienced by French tourists. It focuses on both levels of perceived importance attached to tourism services provided and to different attitudes of satisfaction with the actual quality of service consumed. It also examines Vietnamese tourism service provider perceptions and satisfaction with the products, services and facilities that they have provided to their guests. Overall, the study attempts to determine whether or not differences in cultural values and rules of behaviour exist between the tourist and the host populations. It examines whether these differences will impact on the interaction, perceptions and levels of satisfaction of the two samples. If this is the case, the marketing implications of such cultural variations may be readily assessed. The following sections provide an overview of the staged methodology that will be employed to achieve the research objectives.

The survey sample methodology has several advantages: large sample sizes representing the population under consideration and the use of scoring and rating to measure the attitudes and opinions. Since the results of measurement are easily summarized and analysed, thus generalizations can be made with a degree of confidence in their reliability and measurement for comparative purposes. However, quantitative approaches also have some shortcomings. Rich information on the subjects may not be obtained due to limited interaction between the researchers and the subjects. In addition, large samples are required to facilitate statistical analysis (Brunt, 1997), and they are costly to accurately collect. Surveys may never be precisely indicative of reality and the researcher relies on respondents to be honest and accurate. This creates a potential for biased results. Surveys may also be intrusive if they are not administered skillfully thereby lowering the response rates.

Qualitative data may be more personal and easily understandable (Brunt, 1997; Veal, 1998; Henderson, 1991). Qualitative methods also allow researchers to get close to informants and to grasp their point of view and vision of the world. In this way, qualitative methods can provide meaningful data from a limited number of individuals regarding their expectations, experiences, behaviour, needs and aspirations. However, the qualitative approach has some limitations. "Small numbers of people are normally involved, thus generalisations about the population at large cannot be made. The measurement of qualitative material often requires judgements to be made by the researcher and hence questions of objectivity arise" (Brunt, 1997, p. 18).

The quantitative and qualitative approaches differ in significant ways with each having its strengths and weaknesses. According to Henderson (1991) and Veal (1998) the use of a combination of qualitative and quantitative approaches will enable a researcher to achieve better research results. Furthermore, Newman (1997) notes that "by understanding both styles, we will know a broader range of research and can use both in complementary ways" (p.14).

Since the present cross-cultural study involves the comparison of different cultures with hosts, the quantitative approach is most useful and practical despite the cost of data collection and it seemed to be more achievable than a qualitative one. Primary data were gathered largely using a questionnaire survey of the Vietnamese host sample and the French tourist population. Some qualitative data from the semi-interview and personal observation were also collected to supplement and help interpret the quantitative results from the survey. The interview survey questionnaire technique was chosen for a number of reasons:

- To have the capacity to be administered relatively quickly to a large number of tourists and service providers,
- To achieve a higher total response rate and obtain more complete responses,
- To collect the responses in a format that could be standardized, coded and analysed,
- To collect the data in foreign languages, and
- To allow the researcher an opportunity to give feedback to the respondent and enable interaction with tourists whilst exposed to hosts. For this reason, the social interactions between the researcher and respondent during the personal interview help to reveal feelings and emotions regarding different subjects and to increase the likelihood that a response would be given to all items on the questionnaire.

Sampling is a problem for all research and particularly in the case of cross-cultural research since this approach confronts the extra problem of finding comparable representative subjects (Segall, 1979). Van Raaij (1978) and Segall (1979) have outlined the best types of sampling for each type of cross-cultural research. Random sampling is considered to be most suitable for the comparison of fixed variables such as income, age, education distribution. An attempt was also made to choose respondents from a wide variety of socio-demographic backgrounds of different genders and ages and various social, educational and professional backgrounds. This was done to ensure the representativeness of the samples in terms of the central tendency of their culture. Random sampling is particularly suitable for carrying out statistical tests for the significance of difference. Random sample representativeness is suitable for descriptive studies where the focus of the study is the relationship between variables in different cultures. In this case it is essential that the sample is representative of the cultures that it represents.

A stratified convenience random sampling technique has been used for the purposes of the current study. A probability rather than non-probability sampling design was chosen in the form of a stratified random sample. Ryan (1995) reported that "the stratified random sample does possess several advantages for people who are planning research and using field researchers" (pp. 171-174). By using a stratified sample, the sample adequately reflects the population on the basis of the stratification criteria, namely cultural differences. The total population of French guests (or French tourists) and the Vietnamese hosts were divided into mutually exclusive and exhaustive stratas. As the aim of the study is to illustrate whether or not culture has an affect on perception and satisfaction levels, respondents have been separated into groups based on cultural background or the culture of their spoken language. Western and Asian language groups have been chosen for this study. The French guests represent the Western language group. The Vietnamese hosts represent the Asian language groups. The stratified sample is required to be large, and, to some extent, convenience of sample selection is required to obtain the maximum number of respondents from different language groups, rather than achieving sample proportionality to the total population.

The target population from which the sample has been drawn consists of visitors to Vietnam from France and the Vietnamese hosts (or Vietnamese service providers). Subjects from both the tourist and host populations were assured that the survey was anonymous, confidential and voluntary under the guidelines required by ethics standards for Victorian University. The first population was drawn from the French residents visiting Vietnam on an organised tour or as free independent travellers and whose journey purpose is a pleasure holiday, and other purposes such as business, conferences, visiting friends and relatives and education. The sample comprises 180 French tourists.

Respondents were approached in a variety of source locations in Vietnam, where there is a large concentration of Western tourists. These places are the most frequently visited by the tourists including major attractions, restaurants, shops, hotels, bars in selected major cities in Vietnam such as Ho Chi Minh City, Danang, Hoi An, Hue, Hai Phong and Hanoi cities. As such the individual sample members were collected as a convenience random sample.

The second relevant population investigated consisted of Vietnamese hosts who were working in the tourism industry. For the purpose of this study, the Vietnamese hosts were drawn from various sectors of tourism including accommodation, food and beverage, retail, transportation, customs and immigration and tour operations. These subjects were involved in providing services, products and activities for consumption or purchase by international tourists. The sample comprises 205 Vietnamese hosts. A stratified random sampling was also chosen for the Vietnamese host population as was the case with the tourist population random sample, on the basis of employment. Samples were selected as a convenience random sample from each stratum.

According to Hair et al. (1998), the sample size plays an important role in the estimation and interpretation of the results and provides a basis for estimation of sample error. Moreover, Kline (1994) also indicated that it is essential that the sample should be sufficiently large to enable factor analysis to be undertaken reliably. Nevertheless, there is no consensus on what the sample size should be. The sample size used in this study satisfies Sudman's (1976) requirement for a minimum size of 100 in each group to be compared. The Vietnamese host sample (205) and the French tourist samples (180) are large enough and can be divided into groups, each with a sample size of more than 100.

The present study is exploratory in nature and quantitative analysis is used. Survey and personal face-to-face interviews were used to achieve a high quality of responses and response rate. The survey was administered during 2003 and 2004 on all days of the week to ensure a genuine cross-section of respondents and included subjects from a range of backgrounds, occupations and ages. Prospective respondents were approached in settings where groups of French tourists tend to congregate. Locations included major attractions, restaurants, shops, hotels and bars in Ho Chi Minh city, Danang, Hoi An, Hue, Hai Phong and Hanoi.

Self-completion questionnaires are widely regarded as generating the most reliable responses, since respondents have the opportunity to review the completed questionnaire or revisit questions that were not initially answered. Once the researchers had identified themselves, respondents were provided with information about the intent and content of the survey. Respondents were assured that the survey was anonymous, confidential and voluntary. All questionnaires were returned, whether complete or incomplete. The interview was designed to solicit a wide range of information.

Instrument Design

To meet the purposes of the present study, two versions of the questionnaire were administered to two groups of respondents - the host and the tourist samples. These items were adopted from Rokeach's Values Survey (1973); Argyle's et al. (1986) study on cross-cultural variations in relationship rules; Parasuraman's et al. (1986) study on service quality dimensions. Most questions were identical for the two samples in order to facilitate comparison. The questionnaire was initially designed in English and subsequently translated into Vietnamese and French language. Finally a professional translator back translated the questionnaire into English for purposes of equivalence. The questionnaire consisted of three parts:

- Part 1: Cultural Values refer to the strong beliefs or opinions of what is important, appropriate and socially acceptable in life in a range of situations. Cultural values influence perceptions, constitute evaluation criteria and determine satisfaction. Cultural values determine rules of behaviour and represent individual attributes that influence perceptions. They can be used to analyze cross-cultural differences between people. In this study, cultural values are chosen because cross-cultural holiday satisfaction can be assessed by identifying similarities and differences between tourist and host cultural values, rules of behaviour, perceptions and satisfaction.

This part of the questionnaire involved the application of Rokeach's Value Survey (RVS) (Rokeach, 1973). The RVS technique has been used to test the cross-cultural research purposes and has been shown to be a viable instrument. This instrument was assessed as the best-known survey for measuring human values, and thanks to its universal theory, it has been widely used in many cultural contexts as a reliable and valid measurement of values (Kamakura & Mazzon, 1991). The RSV is a suitable instrument for the current study as it establishes an effective dimensional framework for understanding cultural differences between French tourists and Vietnamese hosts in the context of holiday perceptions and satisfaction. Using a 6-point scale, respondents were asked to rate 18 terminal and 18 instrumental values, with 1 indicating 'Completely Unimportant' and 6 meaning 'Extremely Important'.

- Part 2: Rules of Behaviour are the principles guiding interpersonal interactions which govern overall standards of behaviour and customs. Because rules of behaviour differ between various cultures, as does the interpretation of their appropriateness in social situations, they are chosen in this study for analyzing the cross-cultural differences between peoples' perceptions and satisfaction. Host-guest interactions are an important part of service delivery, and intercultural encounters of the French tourists and Vietnamese hosts can contribute to tourist overall satisfaction. Encounters may encompass greetings, inquiries, casual conversations and business transactions. They occur when tourists talk to service providers whist on holiday such as in planes, buses, hotels, restaurants, shops or tourist attractions.

This part was adapted from Argyle et al.'s study (1986) on cross-cultural variations in the rules which determine relationships. Argyle et al. (1981) argued that the best way to identify social rules is to ask people to rate their importance when applied in a particular social context. Building on the administration of the original Argyle questionnaire in a Western setting, a number of modifications have been made to suit the purposes of the current study. Items which were irrelevant to respondent social interactions in the tourism context were deleted. Some new items were included, namely rules governing social visitation, family invitations and sexual activity. These involved rules which are integral to Vietnamese culture and include conforming to status, conforming to rules of etiquette, having a sense of shame, avoiding embarrassment and avoiding argument. These items were drawn from the literature on interpersonal relations within Asian cultures.

- Part 3: Perceptions and Satisfaction of Tourism Services. Tourism services consist of services offered directly to tourists by Vietnamese hosts from different sectors. In the current study, tourism services are highlighted by the attributes and performance of the Vietnamese hosts as they are the fundamental components in contributing to tourist satisfaction.

Part 3 was designed to measure service attributes and performance. Many of these items were adapted from Parasuraman's 10 service quality dimensions (1985, 1988) as they covered the most important service quality criteria. The ten dimensions composed of tangibles, reliability, responsiveness, communication, credibility, security, competence, courtesy, understanding, knowing the customer, and access. However, as the SERVQUAL instrument was initially designed for a generic measurement of service quality such as banking, management or general services, it does not adequately account for criteria that contribute to the overall quality of tourism services. For that reason, the original scale of SERVQUAL was modified and supplemented by using additional categories that could measure both Asian and Western perceptions of service quality. Some additional variables were added to the questionnaire such as knowledge of Vietnamese hosts about the French culture and customs, and knowledge of tourists about Vietnamese history and culture. These variables were chosen from the focus group discussions with the Vietnamese tour guides who indicated the need to include those variables when evaluating Asian and Western tourists' perceptions of Vietnamese service providers. It was predicted that these variables would be useful for measuring respondent perceptions of service attributes and performances.

Part 3 was measured on a 6-interval point scale and was composed of two sub-sections:

- Section 3a: comprised twenty-nine items and was designed to investigate levels of importance. Respondents were asked to rate their perceived level of importance of the service attributes and performance on a 6-interval point scale. A value of 1 was assigned to an attribute perceived as the least important Completely Unimportant, and a value of 6 was assigned to an attribute perceived as the most important Extremely Important.
- Section 3b: consisted of the same twenty-nine items as appeared on Section 3a. It was designed to investigate respondent levels of satisfaction with actual experience. Respondents were asked to rate their satisfaction with the service attributes and performance on a 6-interval point scale. A value of 1 was assigned to an attribute perceived as the least satisfied Completely Unsatisfied, and a value of 6 was assigned to an attribute perceived as the most satisfied Extremely Satisfied. Overall satisfaction was determined by the degree of difference between the perceived importance and satisfaction with actual experience on service attributes and performances.

The first draft of the questionnaire was piloted among a group of twenty French tourists who had previously visited Vietnam, and also with twenty Vietnamese hosts. The aim of the pilot study was to determine appropriate questions for measuring the relevant concepts and to assess the reliability and validity of the questionnaire. Subjects were asked to evaluate the questionnaire in terms of meaningfulness, style, clarity and difficulty or ease of completion.

Reliability and Validity

The returned survey instruments were checked for omissions, legibility and consistency. Editing procedures were conducted to make the data ready for coding and transference to data storage. The coding categories were exhaustive, providing for all responses. They were also mutually exclusive and independent so that there was no overlap between categories. The raw data was transferred to computer and analysed using Statistical Packages for the Social Science (SPSS) version 22.

Two data sets were created (one list for a host survey and one for a tourist survey). Although many variables were identical in the tourist and the host surveys, these variables were given distinct names for comparison purposes. Since the questionnaire was structured and coding of variables was planned in advance, the questionnaire had categories that have been already built into the answers. In total, two files, one tourist file and one host file were generated in SPSS package. The host file recorded 205 cases and contained 109 variables for each case. The tourist file recorded 180 cases and contained 116 variables.

The stored data were subjected to final screening for completeness, consistency and accuracy. Univariate descriptive statistics were used to inspect the inputs for accuracy. For instance: a) the range of each variable was checked for out-of-range values; b) frequency counts were performed and the distribution of each variable was analyzed to detect irregular answers, outliers, and cases with extreme values; c) the means and standard deviations were computed.

Acording to Sechrest et al. (1972) an instrument is equivalent across systems to the extent that the results provided by the instrument reliably describe with (nearly) the same validity a particular phenomenon in different social systems. Similarity of factorial structure is recommended by these authors as well as by Hui and Triandis (1985) to assess the structural similarity of a construct across cultures. If a construct is the same in two different cultures, it should have the same internal structure in both cases.

The purpose of the reliability assessment is to check for validity and to improve the quality of the measure. Unreliable measures lead to decreased correlation between measures. If no significant relationship exists between constructs, it is impossible to know whether the result is true or due to the unreliability of the measure (Peter, 1979). Reliability refers to the extent to which a scale produces consistent results if repeated measurements are made (Malhotra, 1996; Peter, 1979). More specifically, Churchill (1979) indicated that: "a measure is reliable to the extent that independent but comparable measures of the same traits or construct of a given object agree" (p. 65).

Most single-item measures have uniqueness or specificity that demonstrates a low correlation within a construct, and little relation to other constructs. Many constructs are too complex to be measured effectively with a single-item scale. In the present study, the use of multi-item measures can overcome the weakness of single-item measures, so that multi-item scales are required to achieve both reliable and valid scales (Peter, 1979). The use of a multi-item measurement scale can average out the specificity during aggregation of the item score. It allows for greater distinctions to be made between groups, compared to the single-item measures used to categorise items into a relatively small number of groups. Churchill (1979) indicated that single items have high measurement error and lower reliability because the measure is unlikely to be checked in sequential use of the measurement items relative to multi-item measurement. Along the same lines, Finn and Keyande (1997) proposed that it is best to use multi-item measures because they exhibit high reliability and validity. Furthermore, Peter (1979) reported that the multi-item measurement scale to tap into a construct is a possible way of improving reliability and decreasing error.

Cronbach alpha or Coefficient alpha (Cronbach, 1951) is the most common method accepted by researchers for assessing the reliability of multi-item measures (Anderson and Weitz, 1990). It is a measure of the internal consistency of a set of items, and is considered "absolutely the first measure" which should be used to assess the reliability of a measurement scale (Churchill, 1979; Nunnally, 1978). A low coefficient alpha indicates that the sample of items does not capture the construct and is not shared in the common core of the construct. Such items should be eliminated in order to increase the alpha. Nunnally (1967) suggests that an acceptable alpha is between 0.50 and 0.60. Bruner and Hensel (1993) suggest an alpha of 0.76 and 0.77. In the present study, alpha is calculated for the major constructs of cultural values, rule of behaviour, perceived importance and satisfaction with actual experience of. All coefficient alphas are at an acceptable level, and range between 0.54 and 0.82.

The validity of a scale is defined as "the extent to which differences in observed scale scores reflect true differences among objects on the characteristic being measured, rather than systematic or random error" (Malhotra, 1996, p. 306). There are three main types of validity: content validity, criterion validity and construct validity. Criterion validity can be classified into predictive and concurrent validity. Construct validity can be further categorised into nomological validity, convergent and discriminate validity. Each of these types is used in assessing the validity of the items in measuring the constructs.

A variety of methods have been used in marketing research to test construct validity. Nomological validity is usually established by testing hypotheses developed from a theoretical framework. Peter (1981) suggests that a high internal consistency established through inter-item correlation (i.e., reliability tests) provides support for construct validity. Factor analysis, correlation, and more advanced analysis procedures including confirmatory factor analysis and path analysis are methods for investigation of convergent and discriminant validity. For example, to test for convergent and discriminant validity Kim and Frazier (1997) used a confirmatory factor model, whereas Heidi and John (1988) used correlation and regression analysis. The objective of construct validity is to demonstrate the validity of the key research constructs. In the current study, factor analysis examines convergent and discriminant validity. The factor scores for the whole sample are considered in the assessment of convergent and discriminant validity. The results show that all of the constructs demonstrate strong convergent validity as the final measures load strongly on one factor, and strong discriminant validity as they load lowly on the others. (see The Principal Component Analysis results).

Methods of Data Analysis

For the current study, the quantitative data was analysed using the SPSS version 22. The Principal Components Analysis (PCA) was conducted in order to: 1) group variables in order to derive principal factors for comparisons; and 2) measure the strength of the relationship between each variable and its associated factor. The Exploratory Factor Analysis (EFA) was conducted to explore several key Cultural Dimensions (Cultural values and Rules of behaviour) and Behavioural Dimensions (Perception and Satisfaction) for the two groups: Vietnamese hosts and French tourist samples.

The literature has revealed two essential assumptions underlying Exploratory Factor Analysis: the importance of a sufficiently large sample size and the appropriate factorability of the data. Although the solution of factor analysis is enhanced if variables are normally distributed, the assumption of normality is not critical. Hair et al. (1995) indicated that normality is only necessary if a statistical test is to be applied to the significance of the factors.

Factor Analysis is based on the determination of correlations between variables. If the correlations are small, the data are inappropriate for factor analysis. According to Norusis (1993), in order to test the factorability of variables, there are three kinds of tests as described below:

- The Kaiser-Meyer-Olkin (KMO) measure for sampling adequacy: compares the magnitudes of the observed correlation coefficients to

the magnitudes of the partial correlation coefficients. If the KMO measure is greater than 0.5, then the factorability is assumed.
- Barlett's Test of Sphericity: tests if the correlation matrix of data is an identity matrix and to find out whether there is a relationship between the variables. If Barlett's test rejects the hypothesis that the correlation matrix is an identity matrix, then factorability is assumed, that is, there exist significant correlations among the variables (items) for Exploratory Factor Analysis (EFA).
- Measure of Sampling Adequacy (MSA) for each individual variable. The MSA has the same meaning as the KMO but is measured for each individual variable instead of the whole matrix. Variables with a MSA measure below the acceptable level of 0.5 should be excluded from factor analysis.

Principal Components Analysis (PCA) was conducted in the present study to extract several key dimensions of the Vietnamese hosts that are comparable with those of the French tourist sample. There are as many components as variables but only the largest are extracted. The first principal components account for the most variance and the components are ordered by size as they are extracted. For the initial factor extraction and for determining the number of factors, the study uses the criterion *Eigenvalue* greater than 1. The *Eigenvalue* of each component indicates how much variance is accounted for in the correlation matrix, and is thereby a measure of relative importance for each principal component. Factors retained that exceed an *Eigenvalue* of one are known to be more readily interpretable than factors with eigenvalues less than or equal to one (Turner, 1991). The rationale for the *Eigenvalue* being greater than 1 is that any individual factor should account for the variance of at least a single variable if it is to be retained for interpretation purposes. It is also a requirement that more than one variable loads significantly on any factor. If this is not the case it is not possible to define the dimension adequately.

Kline (1994) and Cattell (1978) reported that in large matrices, the *Eigenvalue* greater than 1 criterion greatly overestimates the number of factors and may split a major factor into several trivial factors. As a result, the scree test (Cattel, 1978) can also be used after the initial factor extraction to select the correct number of factors for factor rotation. According to Kline (1994), Norusis (1993) and deVellis (1991), Cattell's (1978) scree test is a good solution for selecting the correct number of factors and this test must be performed on PCA (Kline, 1994). In a scree plot, the cut-off point for selecting the correct number of factors is where the line suddenly changes slope. If the slope change begins at the k' factor, then k is the true number of factors. In the present study the scree test was used to identify an appropriate number of factors to be retained by using the cut-off points (sudden change of the slope). Moreover, the cumulative percentages of the variance extracted by factors can also be used to decide the significance of the derived factors. Therefore, the multiple decision criteria were used in this study to determine the number of factors to be retained. According to Kline (1994), it is usual that an initial solution of factor analysis does not make it clear which variables belong to which factors. Factor rotation is used to simplify the factor structures and to make them more interpretable. Gorsuch (1983) reported that the selection of method depends on which rotation results in a simpler, more interpretable resolution. For instance, in the oblique rotation, rotated factors may be correlated to each other but in an orthogonal rotation, rotated factors are uncorrelated. The orthogonal rotation was selected for the purpose of the present study.

There is no specific rule for judging the significance of factor loadings (the correlations of the variables with the factors). According to Hair et al. (1998), if the loadings are 0.5 or greater, they are considered practically significant. Moreover, Comrey in Hair et al. (1995) proposed that loadings in excess of 0.63 (40%) are very good, and above 0.7 (50%) are excellent. Nevertheless, Hair et al., (1998) suggested the accepted loadings can be reduced with large sample size and the number analysed. In the current study, a factor loading of over 0.5 has been applied. The Principal Component Analysis (PCA) was conducted to establish the dimensions of cultural difference and holiday perception and satisfaction with tourism products and services in Vietnam from the hosts and guests perspectives. For instance, it was used to establish the dimensions of differences in two groups -Vietnamese hosts and the French tourist sample - with particular regard to cultural values, rules of behaviour, perceived importance and satisfaction of hosts' products and services.

Chapter 4

RESULTS

The Section identifies the cultural similarities and differences influencing the perception and satisfaction levels of Vietnamese hosts and the French guests towards Vietnamese host service attributes and performance. To achieve this objective, Principal Components Analysis (PCA) is used to determine any conceptual groupings evident in the cultural dimensions of cultural values, rules of behaviour, perceived importance and satisfaction in terms of services received by tourists or offered by service providers. The meanings of the identifiable cultural dimensions were discussed and the major cultural differences between Vietnamese hosts and the tourist samples were presented.

According to Lewis-Beck (1994), factor analysis may be used to explain the complex and poorly defined interrelationships evident amongst large numbers of variables. For the purposes of the present research, a factor analytic technique was used to achieve the following objectives:

1) To reduce the number of variables and to determine the prospects for grouping these variables,
2) To identify any structural aspects of the relationship between variables,
3) To create a set of common underlying factors and to make the factor structure more readily interpretable,

4) To determine how many factors are needed to provide an adequate representation of the data, and
5) To determine the extent to which the samples differ on the basis of a set of dimensions.

Principal Components Analysis (PCA) is a commonly used method for undertaking factor extraction with a view to exploring the interrelationships amongst variables. In the present study, the PCA was used to identify: a) groups of cultural values; b) groups of rules of behaviour, and c) perceived importance and satisfaction with actual experience with service attributes and performance of the Vietnamese hosts and the French tourists.

In the case of all two samples, the matrices for factor analysis report a number of correlations exceeding 0.3. This suggests that the factor analytic model is appropriate. In all two cases, the anti-image correlation matrices are small, indicative of the interrelationships among the variables and the sampling adequacy of each variable. All matrices were judged by the researcher as being suitable for the purposes of factor analysis. In the case of the Vietnamese hosts and the three tourist samples, the correlations between variables are significant at a 0.000 level as determined using Barlett's test of sphericity. The Kaiser-Meyer-Olkin (KMO) measure of sampling adequacy ranges between 0.52 and 0.91. In the case of the Vietnamese service providers, the KMO is between 0.63 and 0.74. For the French sample, the KMO is between 0.64 and 0.82. These values are above the acceptable level of 0.50, indicative of satisfactory factorability on the part of the items. The samples used in the present study are sufficient in size, thereby offering the prospect of producing reliable factors.

The number of factors extracted is vital for rotation purposes since it determines the number of groups of variables defined. The purpose of rotating the factors around the origin is to increase the 'fit' of the factors to groups of variables (Turner, 1991). In the case of the current study, the orthogonal rotation is used to reduce the number of variables to a smaller set of independent factors, regardless of whether the resulting factors are meaningful. The varimax approach is used to reach the maximum possible simplification of the columns of the factor matrices. The objective is to achieve a clearer separation of the factors and to identify the variables most representative of these factors (i.e., those with the highest loadings). For the purposes of interpreting the rotated factors, the present study adopts the position that loadings of 0.50 or above are significant.

The variables are classified into two main groups including *Cultural Dimensions* and *Behavioural Dimensions* with the intention of meeting the objectives of the study. These are:

1) Cultural Dimensions
    - Cultural values
    - Rules of behaviour
2) Behavioural Dimensions
    - Perceived importance of service providers' attributes and performance
    - Satisfaction with actual experience of service provider attributes and performance

The interrelationships between dimensions are also investigated with a view to exploring the similarities and differences of the cultural dimensions and behavioural dimensions between the four samples including the Vietnamese service providers and the French tourists.

COMPARATIVE PCA RESULTS BETWEEN HOSTS AND GUESTS

Cultural Values by Sample

The Vietnamese

For the purposes of measuring cultural values, the Measure of Sampling Adequacy is 0.63 in the case of Vietnamese service providers. This finding provided confirmation that the analysis of particular samples was significant. Based on the unrotated factor solution, 14 factors with Eigenvalues of greater than 1 may be extracted. Four factors (F8, F11, F12 and F14) were eliminated from analysis because their definition and correlation was confined to a single variable. Factors F5, F7 and F13 were defined by variables which were uncorrelated with each other, so they were eliminated due to the problem of their interpretation.

As demonstrated in Table 3, the rotated component matrix indicates that the retention of a seven-factor solution F1, F2, F3, F4, F6, F9 and F10) accounts for 63.20% of the explained variance with the first factor responsible for a 9.80% share. Since the variables load significantly on the factors that are well defined by two or more variables, the seven-factor solution for the 36 cultural values variables in the Vietnamese host sample is considered to be acceptable. The seven factors retained for the purposes of further analysis can be summarized as follows:

- Factor 1: *Acumen & Gratification* consists of variables that describe the cues associated with self-fulfilment such as to be knowledgeable and have an understanding of life, be admired by others, to have close companionship with happiness, affection and tenderness with others.
- Factor 2: *Cheerfulness* consists of variables that reflect logic or common sense, as well as light heartedness and joyfulness.
- Factor 3: *Durability* consists of variables reflecting the importance of having an open-mind, ambition or aspiring to a work ethic as well as living a prosperous or comfortable life.
- Factor 4: *Forgiveness* consists of variables reflecting a world with beautiful nature or arts and a willingness to forgive others.
- Factor 6: *Safety & Security* consists of variables that contribute to a secure life and taking care of family members in a world that is free of war and conflict.
- Factor 9: *Salvation* consists of variables reflecting the importance of having an eternal life (salvation) and a dutiful and respectful attitude towards parents and ancestors (obedience).
- Factor 10: *Quality of Life* consists of variables reflecting the expectation of a stimulating or enjoyable leisurely life.

The French Sample

Within the French sample, 10 factors were found with an Eigenvalue of greater than 1 comprising 62.79% of the explained variance with the first factor responsible for a 9.24% share. The Measure of Sampling Adequacy is 0.64. This confirmed that the analysis of particular samples was significant. Factors F3, F6 and F10 were defined by variables which were uncorrelated with each other, so they were eliminated from further analysis since they could cause interpretation problems. The seven factor solution for the 36 cultural values variables in the French sample (F1, F2, F4, F5, F7, F8 and F9) were retained for further analysis as they were well defined by two or more variables. The seven-factor solution is accepted as reliable. Its structure is identified in Table 3 and may be summarised as follows:

- Factor 1: *Esteem-Personal Contentedness*: refers to variables that describe indicators associated with having freedom and honesty.
- Factor 2: *Salvation & Inner Harmony* consists of variables reflecting the importance of having an Inner Harmony and the search for eternal life (salvation).
- Factor 4: *Competence* refers to personal meaning in life such as being intellectual and ambitious.
- Factor 5: *Sense of Accomplishment* consists of variables that describe the cues associated with self-fulfilment such as having social recognition and loving.
- Factor 7: *Sense of Self* consists of variables reflecting the importance of being logical and having a comfortable life.
- Factor 8: *A World of Beauty* consists of variables reflecting the importance of being polite and the expectation of having a world of beauty.
- Factor 9: *Hedonistic Values* consists of variables reflecting the expectation of having happiness and a stimulating or enjoyable leisurely life.

Table 3. Results of the varimax rotated component matrix for cultural values

Vietnamese Hosts *(N=205)*	LD	French Tourists *(N=180)*	LD
KMO= 0.632		KMO = 0.648	
Bartlett's Test = 1332.819		Bartlett's Test = 826.009	
Sig. = 0.000		*Sig. = 0.000*	
F1: Acumen & Gratification & Friendship		F1: Esteem & Personal Contentment	
Wisdom	0.67	Freedom	0.74
Happiness	0.66	Honest	0.64
True friendship	0.66		
Loving	0.65		
Helpful	0.54		
E%V = 9.79		*E%V = 9.24*	
F2: Cheerfulness		F2: Salvation	
Logical	0.80	Salvation	0.78
Cheerful	0.53	Inner harmony	0.58
E%V = 7.21		*E%V= 8.25*	
F3: Durability		F4 : Competence	
Broaded-minded	0.78	Ambitious	0.72
Ambitious	0.62	Intellectual	0.63
A Comfortable Life	0.60		
E%V = 6.92		*E%V = 6.23*	
F4: Forgiveness		F5: Sense of Accomplishment	
A World of Beauty	0.69	Social recognition	0.74
Forgiving	0.58	Loving	0.62
E%V = 6.11		*E%V = 5.97*	
F6: Safety & Security		F7: Sense of Self	
Family security	0.67	A Comfortable Life	0.76
A World of Peace	0.66	Logical	0.51
E%V = 5.77		*E%V = 5.77*	
F9: Salvation & Obedience		F8: A World of Beauty	
Salvation	0.79	A World of Beauty	0.74
Obedient	0.53	Polite	0.66
E%V = 4.97		*E%V = 5.67*	
F10: Quality of Life		F9: Hedonistic Values	
An Exciting Life	0.75	Happiness	0.74
Pleasure	0.53	Pleasure	0.52
E%V = 4.50		*E%V = 5.48*	

LD: Factor Loading;
E%V: Explained Percentage of Variance
KMO: Kaiser-Meyer-Olkin Measure of Sampling Adequacy
Bartlett's Test: Bartlett's Test of Sphericity
Sig.: Significance.

Rules of Behaviour by Sample

The Vietnamese

Twelve factors with Eigenvalues of greater than 1 may be extracted based on the unrotated factor solution. Since their definition and correlation were confined to a single variable, two factors were eliminated from analysis (F9 and F10). Factors F2, F5, F6 and F12 were defined by variables which were uncorrelated with each other. These factors were eliminated from further analysis since they might cause interpretation problems.

The rotated component matrix indicates that the six-factor solution (F1, F3, F4, F7, F8 and F11) accounts for 63% of the explained variance, with the first factor responsible for a 13.1% share. As demonstrated in Table 4, the dimensions consist of variables with significant factor loadings of 0.5 and above. Since the variables load significantly on the factors that are well defined by two or more variables, the six-factor solution for the 34 Rules of behaviour variables applicable to the Vietnamese sample is acceptable. The six factors retained for further analysis are as follows:

- Factor 1: *Sense of Hierarchy* refers to variables that reflect the importance of conforming to the principles of social stratification such as rules of etiquette and obeying instructions. Conformity also implies being neatly dressed, avoiding complaining, showing respect for others, having a sense of shame and acknowledging the birthday of others.
- Factor 3: *Social Relations* refers to variables that describe indicators associated with establishing a relationship with others by seeking material assistance or by asking personal questions.
- Factor 4: *Consideration for Others* consists of variables that describe the cues associated with paying attention to demonstrating selflessness towards others. This involves showing an interest in others and asking for personal advice.
- Factor 7: *Sense of Obligation* consists of variables that reflect the responsibility for commitment to others such as seeking a chance to repay favours, or using appropriate titles when addressing others.

- Factor 8: *Pursuit of Social Harmony* consists of variables reflecting the importance of avoiding arguments or discussion about sensitive issues in order to maintain social harmony
- Factor 11: *Communication Behaviour* refers to variables that describe signals associated with verbal and non-verbal behaviour, such as maintaining eye contact during conversation, and by avoiding making fun of others.

The French

Based on the unrotated factor solution, 12 factors with Eigenvalues of greater than 1 may be extracted. Four factors (F1, F3, F7 and F12) were defined by variables which were uncorrelated with each other. These variables were eliminated since they could result in interpretative problems.

The rotated component matrix indicates that the eight-factor solution (F2, F4, F5, F6, F8, F9, F10 and F11) accounts for 64% of the explained variance, with the first factor responsible for a 16.1% share. As demonstrated in Table 4, the dimensions consist of variables with significant factor loadings of 0.5 and above. Since the variables load significantly on the factors that are well defined by two or more variables, the eight-factor solution for the 34 Rules of Behaviour variables in the French Sample is acceptable. The eight factors retained for further analysis are as follows:

- Factor 2: *Interest in Others* consists of variables associated with demonstrating consideration for others. This involves showing respect and affection for others in public and conforming to the status of others.
- Factor 4: *Respect for Privacy* consists of variables that reflect the importance of respecting privacy. This entails taking time to develop relationships, respecting the privacy of others and avoiding request by others for personal advice.
- Factor 5: *Clarity of Expression* refers to variables that describe indicators associated with clearly demonstrating intention in the course of communication and avoiding making fun of others.

Table 4. Results of the varimax rotated component matrix for rules of behavior

Vietnamese Hosts (N=205)	LD	French Tourists (N=180)	LD
KMO= 0.761		KMO = 0.676	
Bartlett's Test = 1639.673		Bartlett's Test =1433.371	
Sig. = 0.000		Sig. = 0.000	
F1: Sense of Hierarchy		F2: Interest in Others	
Avoid Complaining	0.76	Show Interest in Other person	0.78
Be Neatly Dressed	0.76	Show Respect to Other Person	0.61
Conform to Rules of Etiquette	0.74	Show Affection for Other Person in Public	0.61
Show Respect to Other Person	0.69	Conform to Status of Other Person	
Acknowledge Other Person's Birthday	-0.68		
Obey Instructions of Other Person	0.65		
Have a Sense of Shame	0.52		
E%V = 13.11		E%V = 6.71	
F3: Social Relations		F4: Respect for Privacy	
Ask Other Person for Material Help	0.71	Respect other Person's Privacy	0.75
Ask Personal Questions of Other People	0.65	Take Time to Develop Relationships	0.72
		Ask Other Person for Personal Advice	-0.52
E%V = 5.27		E%V = 6.24	
F4: Consideration for Others		F5: Clarity of Expression	
Ask Other Person for Personal Advice	0.76	Indicate your Intentions Clearly	0.79
Show Interest in Other person	0.57	Avoid Making Fun of Other Person	0.63
E%V = 5.24		E%V =	5.23
F7: Sense of Obligation		F6: Display of Feelings	
Address By First Name	0.79	Address By First Name	0.73
Seek to Repay Favours	0.50	Show Emotion in Public	0.50
E%V = 4.43		E%V = 4.82	
F8: The Pursuit of Social Harmony		F8: Compliment	

Table 4. (Continued)

Vietnamese Hosts (N=205)	LD	French Tourists (N=180)	LD
Avoid Arguments	0.72	Compliment Other Person	0.70
Talk about Sensitive Issues	-0.50	Swear in Public	-0.51
E%V = 4.26		E%V = 4.64	
F11: Communication Behaviour		F9: Individual Needs	
Look in the eyes during Conversation	0.77	Think about Own Needs	0.74
Avoid Making Fun of Other Person	0.57	Avoid Embarrassment	-0.72
E%V = 4.10		E%V = 4.62	
		F10: Form of Greeting	
		Be Neatly Dressed	0.67
		Shake Hands with One Another	0.65
		E%V = 4.50	
		F11: Communication Behaviour	
		Express Personal Opinions	0.74
		Look in the eyes during Conversation	0.50
		E%V = 4.40	

LD: Factor Loading
E%V: Explained Percentage of Variance
KMO: Kaiser-Meyer-Olkin Measure of Sampling Adequacy
Bartlett's Test: Bartlett's Test of Sphericity
Sig.: Significance.

- Factor 6: *Display of Feelings* consists of variables reflecting the importance of showing emotion in public or addressing other people by their first name.
- Factor 8: *Compliment* refers to variables that describe the cues associated with expressing admiration and for not swearing in public.
- Factor 9: *Individual Needs* consists of variables reflecting independence, an orientation towards others, attaching importance to personal needs and avoiding causing embarrassment to others.

- Factor 10: *Form of Greeting* consists of variables reflecting the importance of shaking hands with one another and being neatly dressed when with others.
- Factor 11: *Communication Behaviour* refers to variables that describe the cues associated with expressing personal opinions and looking in the other's eyes during conversation.

Behavioural Dimensions by Sample

What differences are evident amongst Vietnamese hosts and the French tourists in the levels of perceived importance and ultimate satisfaction with service attributes and performance? Principal Components Analysis (PCA) has been used to answer these questions. In the current study, tourist satisfaction towards host service attributes and performance was measured by comparing the differences between the levels of importance of service and actual experience of services. In order to reduce the measures of importance and actual services experiences into dimensions capable of summarising each set of measures, the Principal Components Analysis (PCA) was run for each culture and for each data matrix in terms of importance and actual experience. Table 5 presents the generalised results of the PCA for the importance measures. Table 6 presents the actual experience measures. Only variables with loadings exceeding 0.5 are recorded.

Perceived Importance of Service Attributes and Performance

The Vietnamese
For the Vietnamese sample, the Kaiser-Meyer-Olkin Measure of Sampling Adequacy resulted in a figure of 0.745. This provided assurance that the analysis was significant for the given sample. Nine factors had an Eigenvalue of greater than 1, comprising 57.10% of explained variance. Within the nine factors, four (F6, F7, F8 and F9) were eliminated from

consideration because an insufficient number of variables could cause problems for interpretation purposes. The five-factor solution (F1, F2, F3, F4 and F6) for the 29 perceptions of service variables in the Vietnamese sample was retained for further analysis since they were well defined and drew upon two or more variables. The five-factor solution identified in Table 5 may be summarized as follows:

- Factor 1: *Attentiveness* reflects an expectation of the capacity of hosts to behave towards tourists in a way indicative of their concern for the welfare of tourists by offering individualised attention. It also involves a capacity amongst hosts to be responsive to tourist requests such as giving adequate explanations, being approachable and easy to find when needed.
- Factor 2: *Responsiveness* involves the ability of the host to respond promptly to tourist requests and implies the need to handle tourist queries punctually. It also indicates the importance of being punctual in performing services, being a good listener and providing accurate information.
- Factor 3: *Confidence & Trustworthiness* involves hosts being confident in their service performance, behaving in a trustworthy manner and keeping tourists informed.
- Factor 4: *Smart & Polite* indicates the need to relate to tangible cues associated with service such as physical appearance (e.g. neatly dressed) and behaving politely.
- Factor 5: *Intercultural Competence* relates to the host ability to communicate with guests in the required languages. In the case of the French tourist market, the language is French respectively. It is also critical for hosts to be acquainted with European cultures and customs in order to understand the needs of individual tourists.

The French

For the French sample, the Kaiser-Meyer-Olkin Measure of Sampling Adequacy resulted in a figure of 0.616. This provided assurance that the analysis was significant for the given sample. Eleven factors have been extracted with an Eigenvalue greater than 1, comprising 61.40% of explained variance. Factor F10 was defined by variables which were uncorrelated with each other; therefore it was eliminated from further analysis since it could cause interpretation problems. Four factors were eliminated from analysis (F7, F8, F9 and F11) because they were only defined with one variable. The six-factor solution (F1, F2, F3, F4, F5 and F6) for the 29 perceptions of service variables in the French sample were retained for further analysis. The six-factor solution identified in Table 5 can be summarized as follows:

- Factor 1: *Courtesy & Friendliness* reflects the tourist expectation that hosts are approachable and friendly.
- Factor 2: *Verbal Communication* refers to the tourist expectation that hosts are polite, speak French and provide tourists with adequate explanations.
- Factor 3: *Helpful & Trustworthy* reflects tourist expectations that hosts are helpful, respectful and trustworthy.
- Factor 4: *Prompt & Accurate Information* relates to the ability of hosts to solve problems quickly, to provide accurate information and to answer all tourists' questions.
- Factor 5: *Intercultural Competence & Personalized Attention* relates to the ability of hosts to be knowledgeable about French culture and customs in order to offer individualized attention to tourists and be available and easy to find when needed by tourists.
- Factor 6: *Capability* refers to the tourist expectations that hosts are able to perform service and to keep tourists informed.

Table 5. Results of the varimax rotated component matrix for the dimension of importance of service attributes and performance

Vietnamese Hosts (N=205)	LD	French Tourists (N=180)	LD
KMO = 0745		KMO = 0.616	
Bartlett's Test = 1195.659		Bartlett's Test = 689.543	
Sig. = 0.000		Sig. = 0.000	
F1: Attentiveness		F1: Courtesy & Friendliness	
Concerned about Tourists' Welfare	0.63	Approachable	0.78
Responsive to Tourists' Needs	0.60	Friendly	0.75
Give Adequate Explanations to Tourists	0.58		
Easy to find when needed	0.53		
Approachable	0.50		
E%V = 10.40		E%V = 7.47	
F2: Responsiveness		F2: Verbal Communication	
Solve Problems Quickly	0.73	Speak French Language	0.76
Provide Prompt Service	0.73	Polite	0.57
Punctual-Perform Services On Time	0.71	Give Adequate Explanations to Tourists	0.51
Provide Accurate Information	0.50		
Listen to Tourists	0.50		
E%V = 8.82		E%V = 6.52	
F3: Confidence & Trustworthiness		F3: Helpful & Trustworthiness	
Confident	0.75	Helpful	0.67
Trustworthy	0.52	Respectful	0.59
Keep Tourists Informed	0.51	Trustworthy	0.57
E%V = 7.03		E%V = 6.28	
F4: Smart & Polite		F4: Accurate Information	
Neatly Dressed	0.67	Solve Problems Quickly	0.69
Polite	0.63	Provide Accurate Information	0.63
		Answer all Questions	0.57
E%V = 5.70		E%V = 5.84	
F5: Intercultural Competence		F5: Intercultural Competence & Personalized Attention	

Vietnamese Hosts (N=205)	LD	French Tourists (N=180)	LD
Speak English, French, Chinese Languages	0.66	Know Western Culture and Customs	0.73
Know Asian and Western Culture and Customs	0.64	Easy to find when needed	0.62
Understand Western and Asian Tourists' Needs	0.60	Offer Individualized Attention to Tourists	0.51
E%V = 5.68		E%V =5.83	
		F6: Capability	
		Keep Tourists Informed	0.72
		Capable of Performing Service	0.54
		E%V =5.40	

LD: Factor Loading
E%V: Explained Percentage of Variance
KMO: Kaiser-Meyer-Olkin Measure of Sampling Adequacy
Bartlett's Test: Bartlett's Test of Sphericity; Sig.: Significance.

Satisfaction with Service Attributes and Performance

The Vietnamese

For the Vietnamese sample, the Kaiser-Meyer-Olkin Measure of Sampling Adequacy gave a figure of 0.768. This provided assurance that the analysis was significant for the given sample. Nine factors had an Eigenvalue greater than 1, comprising 62.60% of explained variance. Within the nine factors, three (F7, F8 and F9) were eliminated from analysis because the insufficiency of the number of variables could cause problems at the interpretation stage. The six-factor solution (F1, F2, F3, F4, F5 and F6) for the 29 satisfaction with service variables in the Vietnamese sample was retained for further analysis as they were well defined and drew upon two or more variables. The six-factor solution identified in Table 6 may be summarized as follows:

- Factor 1: *Attentiveness* reflects hosts' satisfaction with their ability to behave towards tourists in a way that could indicate their concern for the welfare of tourists. It also entails capacity amongst the hosts to be responsive to tourist requests such as giving adequate

explanations, being approachable, trustworthy and easy to find when needed.
- Factor 2: *Intercultural Competence* reflects hosts'satisfaction with their ability to be acquainted with Asian and Western culture and customs in order to understand and anticipate the needs of the individual tourist, as well as their ability to communicate with guests in the required languages.
- Factor 3: *Responsiveness* reflects hosts' satisfaction with their capability to provide a prompt and punctual service and to solve problems quickly.
- Factor 4: *Smart & Well Mannered* reflects hosts' satisfaction with their polite, respectful and helpful conduct towards their guests. It also refers to hosts' tangible cues associated with their service such as smart physical appearance.
- Factor 5: *Treat Tourists as Guests* reflects hosts' satisfaction with their ability to have good knowledge of Vietnamese history and culture. It also entails host attentiveness and special care in treating their customers as guests.
- Factor 6: *Confident & Considerate* reflects hosts' satisfaction of being confident, considerate and keeping tourists informed.

The French

For the French sample, the Kaiser-Meyer-Olkin Measure of Sampling Adequacy resulted in a figure of 0.600. This provided assurance that the analysis was significant for the given sample. Ten factors have been extracted with an Eigenvalue greater than 1, comprising 63.50% of explained variance. Factor F6 was eliminated from analysis because the insufficiency of the number of variables could cause problems at the interpretation stage.

The nine-factor solution (F1, F2, F3, F4, F5, F7, F8, F9 and F10) for the 29 satisfaction with service variables in the French sample was retained for further analysis. The nine-factor solution identified in Table 6 can be summarized as follows:

- Factor 1: *Courtesy & Friendliness* reflects guest satisfaction with the approachable and friendly manner of hosts.
- Factor 2: *Helpful & Trustworthy* reflects guest satisfaction with the helpful, respectful and trustworthy manner of hosts.
- Factor 3: *Communicative* reflects guest satisfaction with the communicative attitude of hosts by talking and finding hosts easily when needed.
- Factor 4: *Accurate Information* reflects guest satisfaction with the ability of hosts to solve problems quickly and to provide accurate information.
- Factor 5: *Capability* reflects guest satisfaction with the ability of hosts to keep tourists informed and to perform service.
- Factor 7: *Personalized Attention* reflects guest satisfaction with the ability of hosts to be considerate and offer individualized attention to tourists.
- Factor 8: *Verbal Communication* reflects guest satisfaction with the ability of hosts to speak French and to be polite.
- Factor 9: *Confident & Anticipate* reflects guest satisfaction with the confident manner of hosts and the ability of hosts to anticipate the needs of Western tourists.
- Factor 10: *Responsive & Prompt Service* reflects guest satisfaction with hosts' responsiveness and hosts' ability to provide prompt service.

In summary, the Principal Components Analysis (PCA) was used to determine whether or not the basic factoral structure of cultural differences between the Vietnamese hosts and the French tourists existed and was significant. The factorial analysis was applied only for variables which differed between the Vietnamese and French sample. The analysis identified a clear number of dimensions (groupings) of cultural values, rules of behaviour, perceptions of and satisfaction with service components that differed between the Vietnamese hosts and the French guests. The analysis also identified a clear number of cultural dimensions of French language group that differed from the Vietnamese sample.

Table 6. Results of the varimax rotated component matrix for the dimension of satisfaction with service attributes and performance

Vietnamese Hosts (N=205)	LD	French Tourists (N=180)	LD
KMO = 0.768		KMO = 0.600	
Bartlett's Test = 1557.762		Bartlett's Test = 656.896	
Sig. = 0.000		Sig. = 0.000	
F1: Attentiveness		F1: Courtesy & Friendliness	
Give Adequate Explanations to Tourists	0.72	Friendly	0.77
Easy to find when needed	0.63	Approachable	0.76
Trustworthy	0.61		
Offer Individualized Attention to Tourists	0.59		
Concerned about Tourists' Welfare	0.54		
Responsive to Tourists' Needs	0.53		
Approachable	0.52		
E%V = 12.10		E%V = 7.23	
F2: Intercultural Competence		F2: Helpful & Trustworthiness	
Know Asian and Western Culture and Customs	0.74	Helpful	0.71
Understand Western and Asian Tourists' Needs	0.74	Respectful	0.66
Speak English, French, Chinese Languages	0.71	Trustworthy	0.61
Anticipate Western and Asian Tourists' Needs	0.69		
E%V = 8.78		E%V = 6.72	
F3: Responsiveness		F3: Communicative	
Provide Prompt Service	0.79	Easy to talk to Tourists	0.77
Solve Problems Quickly	0.74	Easy to find when needed	0.54
Punctual-Perform Services On Time	0.68		
Capable of Performing Service	0.50		
E%V = 8.76		E%V = 6.52	
F4: Smart & Polite		F4: Accurate Information	
Polite	0.73	Provide Accurate Information	-0.72
Neatly Dressed	0.66	Give Adequate Explanations to Tourists	0.71

//
Vietnamese Hosts (N=205)	LD	French Tourists (N=180)	LD
Respectful	0.56	Solve Problems Quickly	0.69
Helpful	0.53		
E%V = 7.82		E%V = 6.12	
F5: Treat Tourists as Guests		F5: Capability	
Treat Tourists as Guests	0.66	Keep Tourists Informed	-0.74
Know Vietnamese History and Culture	0.58	Capable of Performing Service	0.55
E%V = 5.64		E%V = 5.65	
F6: Confident & Considerate		F7: Personalized Attention	
Confident	0.73	Offer Individualized Attention to Tourists	-0.75
Keep Tourists Informed	0.57	Considerate	0.51
Considerate	0.54		
E%V = 5.61		E%V = 5.49	
		F8: Verbal Communication	
		Speak French Language	-0.71
		Polite	0.61
		E%V = 5.11	
		F9: Confident & Anticipate	
		Confident	-0.66
		Anticipate Western Tourists' Needs	-0.63
		E%V = 4.91	
		F10: Responsive & Prompt Service	
		Responsive to Tourists' Needs	0.61
		Provide Prompt Service	-0.61
		E%V = 4.88	

LD: Factor Loading
E%V: Explained Percentage of Variance
KMO: Kaiser-Meyer-Olkin Measure of Sampling Adequacy
Bartlett's Test: Bartlett's Test of Sphericity
Sig.: Significance.

Chapter 5

INTERPRETATION AND DISCUSSION

COMPARATIVE CULTURAL VALUES AND RULES OF BEHAVIOUR

This study has focussed on Eastern (Vietnamese) and Western (French) cultures. The French society is commonly regarded as being more egalitarian. Given such differences, certain rules of behaviour that are applicable to Vietnamese society may not be shared or even understood by French.

The Vietnamese generally adhere to traditional ways of behaviour and to formal etiquette. Notable features include forms of social order, deference to tradition and authority, and the obligation within and outside the relevant kin group. These rules are designed to prevent damage of reputation, conflict and loss of face in order to maintain social and family harmony in a hierarchical society. In Western societies, formal behaviour is generally unacceptable because prevailing behaviour is casual and less dependent on social position and age. Prominent attributes of the French culture include forms of egalitarianism with an emphasis on independence and self-reliance. Feelings of duty and obligation may be less prominent and rules of formal etiquette more limited in these societies. French people generally conduct their lives according to other principles and manners. The etiquette of social behaviour plays a relatively minor role in their lives. People interact with others on the basis of informality, privacy, flexibility and standing out. People are explicit and display their emotions publicly. The Vietnamese are more implicit and suppress their emotions. The former are more risk-taking and focus on promptness and getting the best deal, but are less concerned about intra-group social harmony.

The French focus on *Egalitarianism*. People are more casual and their behaviour depends less on social position and age. Respect and social recognition in these societies are gained through knowledge, achievement and hard work. By contrast, certain principles of social stratification and age grading within the Vietnamese society may not be readily understood by French. Vietnamese attach great importance to reverence and to *Sense of Hierarchy*. A complex system of grading on the basis of age, occupation, and position within the Vietnamese society requires correct behaviour and respect to be shown towards all those of higher social standing. Within this system of deference to hierarchical authority, customary laws and standards specify how each community member should react to others and the nature and forms of obligations within and outside kin groups. Each age group must, for example, be addressed by the correct terminology and language. Respect is also shown to several other objects of everyday life. These include books, since books bring knowledge and understanding. Rice is respected because it gives life both for the individual and the nation, so that planting and harvesting requires special ceremonies. The Buddha image is most respected by the Vietnamese. This reference helps the Vietnamese to become closer to the spirit of Buddha, to overcome their hardships and to eliminate suffering.

The French conform to the status of others through their social behaviour in life and at work. Though the French attach less importance to hierarchy than their Vietnamese counterpart, they conform to the status of others and have to some degree and rated this dimension as important. A country with large Power Distance and strong Uncertainty Avoidance such as France is more likely to prefer organizations to be top down with clear policies, rules, and regulations, much like a bureaucratic pyramid. The French tend to resolve problems with reference to the hierarchy. The implicit model of a well-functioning organization for the French is a pyramid (Hofstede, 1980). An appropriate analogy refers to a French doctoral student (pyramid) who is allocated a Dutch professor as supervisor (Dutchviwed viwed as a village market). The student waits for the professor to provide guidelines for the thesis, while the professor waits for the student to approach him or her to discuss the project. In French society, rules and regulations are in place to achieve predictability within the organisation. French managers see their organisations not only as political systems but also as authority systems, as role formalisation systems and hierarchical relationship systems (Laurent, 1983). They have a clear notion of the organisational structure. Hierarchy and power are important and these are legitimised by high-level qualifications (Barsoux and Laurence, 1990a).

Trompenaars and Hampden-Turner (1997) reported that achievement cultures (such as those encountered in the West) ascribe status on the basis of what people do and their "track record." However, ascription cultures (such as Eastern culture) ascribe status on the basis of who people are, including place of birth, kinship, gender, age, class, connections, education and profession. People status is formed not from what they studied, but where and with whom. Vietnamese respondents to the survey attached great importance to *Etiquette*. They conform well to the rules of etiquette and to the status of others. Expressions of etiquette include being polite and obedient, showing respect to others and being neatly dressed. In social relations, the Vietnamese are very formal about matters of etiquette. The common people in Vietnam have a tradition of showing deep respect for the wise, honest and virtuous, irrespective of their position in society. For Vietnamese respondents, introduction is a very important etiquette. They are suspicious of anyone who approaches them without an introduction. They are however willing to accept foreigners without introduction as a gesture of goodwill and politeness. If an introduction emanates from a friend, a superior, or an important business person the Vietnamese always take care of the person. This is not the case for French who can ignore the person introduced without damaging the relationship with the person who provided the introduction.

Trompenaars and Hampden-Turner (1997) reported that individualist cultures elevate individual interests over the group, whereas communitarian cultures submerge personal interests in favour of the group. Reflective of more inner-directed Western society, the French learn to control nature, environment and circumstances. Reflective of outer-directed Asian societies, the Vietnamese learn to harmonize with nature and their environment, seeing them as more powerful than the individual. Higher value is attached to *Pursuit of Social Harmony*. Harmony is found in the maintenance of an individual's face, meaning dignity, self-respect, and prestige. *Social relations* should be conducted to protect the dignity of all. Paying respect to someone else is called 'giving face' (Bond and Hwang, 1986; Redding, 1990). Virtuous behaviour towards others involves treating others as one would like to be treated oneself (Hofstede and Bond, 1988). This type of *Harmony* implies that all *Social Relations* should be smooth, relaxed, pleasant and conflict-free. Harmonious social relations are achieved through being non-assertive, polite, courteous, humble and well-mannered. Moreover, these rules of behaviour are also demonstrated between group members by avoiding complaint, arguing or to making fun of others and even by having a sense of shame about doing something to embarrass or hurt others. Although there are many prescribed rules of social behaviour within Vietnam's social classes, collaterality in interpersonal relations is also maintained, with an emphasis on equality and communal consensus.

Personal relationships between people and in business should be smooth and flow gracefully and unhurriedly. Personal relationships are an important aspect of life. Smooth *Social Relations* require consideration, taking into account the feelings of others, avoiding inconvenience for others or imposing one's own will. Social stability is based on *Wu Lun* (unequal relationship between people). Such relationships are based on mutual and complementary obligations: the junior partner owes the senior respect and obedience; the senior owes the junior partner protection and consideration (Lau and Kuan, 1988). The family is the prototype of all social organizations. A person is not primarily an individual; rather, he or she is a member of a family. Children should learn to restrain themselves and to overcome their individuality in order to maintain the harmony in the family

(Ko et al., 1990). Moreover, care and consideration in *Social Relations* are not confined to the spiritual and sentimental side but sometimes extend to the material level. People may ask personal questions of others or ask others for material help. Such relationships scarcely exist in Western societies, where personal privacy is respected and protected.

The Vietnamese attach great importance to *Consideration for Others*. In contrast to their French counterpart, they demonstrate consideration for the feelings of others openly by being obedient, humble, polite and respectful in order to make people like them and be nice to them. Considering and respecting the feelings of others is related to the hierarchical Vietnamese system of status. Rules of respect are evident between children and parents, between young and old and between teachers and students. Humility and respect is often shown by excusing themselves and not trying to be higher than the other person. This is something that French might find difficult to hide. For the Vietnamese consideration is demonstrated by the interest that they show in others when asking for personal advice.

Westerners emphasise "doing one's own thing" and are less concerned about the consequences of their behaviour on others. It is interesting however to note that French respondents rated the *Interest in Others* dimension as most important. This implies showing interest in and respect for others, showing affection for and complimenting others. These results demonstrate that French respondents and prospectively the French generally have a strong inclination towards affective culture where the expression of human emotions is elevated.

In French society, almost everyone is perceived to be equal and the traditional privileges of royalty, class and subordination are basically rejected (Engel et al., 1995; Harris and Moran, 1996). French are characterized by informality and social reciprocities within social relationships are defined loosely. People acknowledge the *Birthdays of others*. It is customary in French culture to celebrate the anniversary of one's birthday in some way, for example by having a party with friends in which gifts are exchanged. It is also customary to treat someone especially nicely on their birthday and to accede to their wishes.

The Vietnamese have a propensity for celebrating the dead rather than the Birthday. The Vietnamese quietly accept the approach of death as the beginning of a new existence. When they feel that death is near, they spend most of their time preparing for their passage into the spirit world. Many elderly persons order their coffin in advance and select the location for their "last resting place." Elderly persons describe the time of death as "the time to come back to the source, to the origin." This has a parallel with a man who following a lengthy absence from his fatherland, returns joyfully and cheerfully. Death has the name "time to go to a better existence," and is worth celebrating. This explains why the Vietnamese use white for mourning _white is bright and not gloomy and this is a special Vietnamese concept. Nonetheless, thanks to previous exposure to French culture and recent interactions with other Western cultures during the period of economic reforms, the Vietnamese now give greater acknowledgement to birthdays. This is however more characteristic amongst the younger generation.

In Asian cultures, relationships amongst people are very strong. The Vietnamese attach considerable importance to *Sense of Obligation*. They value interdependence, and in particular the concepts of Guanxi (or, social networks) and *Renqing* and *Bao* (or, reciprocation of favours and gifts). These concepts highlight the need for continuity so that affinity for each other is well-established. There is a particular emphasis on long-term, asymmetrical reciprocation in exchange relationships (Cheng, 1988). It is not only important to maintain contact with acquaintances in one's personal *Guanxi* network by greetings, visitations and exchanging gifts. It also involves sympathising and helping other in-group members who are needy, and showing them *Renqing*. *Renqing* and *Bao* should be reciprocated as soon as possible.

In Vietnamese society, special emphasis is placed on duty and family moral obligations, with the family as the basic societal unit. The individual becomes part of the group, personal interests are related through the group, and all interactions depend upon other group members. Filial piety is the basis of Vietnamese morality and ethics reflective of the relationship of parent and child and the moral obligation of the child towards the parents.

The parents provide for the child and the child must provide for the parent in return. For this reason, a *Sense of Obligation* and seeking a chance to repay favours to the parents or other family's members is regarded as an essential attribute in Vietnamese society. In daily life, the *Sense of Obligation* is well illustrated, not only by the manner with which people try to help and care for each other, but also how they seek opportunities to pay their favours to others. *Guanxi*, a symbolic trait of Vietnamese culture is a common social relationship which is widely recognised and established in every kind of relationship within Vietnam including in business. Within Vietnamese society, Guanxi refers to the status and intensity of an ongoing relationship between two people, which extends to others who are part of the social networks of the two individuals (Kirkbride et al., 1991, p. 370). According to Trompenaars and Hampden-Turner (1997), communication occurs differently in affective and neutral cultures. This leads to different communication styles between Western and Eastern cultures. The Latin oriented cultures such as French interrupt each other to show how interested they are in what is being said. In the case of Orientals (eg. the Vietnamese), a moment of silence is to show respect to the other or by processing what was said before speaking. Consequently, disparities are evident towards communication behaviour. In French society, non-verbal cues are expressed explicitly clearly through *Communication Behaviour*. People are generally friendly, open and straightforward. They attach importance to *Clarity of Expression* and indicate their intentions clearly in the course of their communication. In Western societies an intentional expression of personal opinion, or looking directly into the eyes of another person during conversation, are very encouraging because they are the signs of honesty and respect. When communicating with others, the message is primarily in the spoken language, and is not overridden by non-verbal communication such as smiles, gestures, eye contact and silence. For French, a credible person has to be articulate and outspoken while being direct, rational, decisive and unyielding. By contrast, avoiding looking into another person's eyes when talking or maintaining a 'monotonous' voice is a way of showing self-control and respect in Oriental societies. Vietnamese hosts rated *Communication Behaviour* as an important issue. Even though this dimension was ranked differently, the Vietnamese emphasis on this attribute had similarities with the French response. The older generation of Vietnamese people maintain the traditional attitude of avoiding looking into the eyes of another when talking as a sign of respect. The present study arrived at a conclusion which differs from the widely held view. This may be because most Vietnamese respondents were generally younger and relatively more exposed to Western culture and values through experience of service encounters with tourists.

Neutral cultures elevate objectivity over emotional expression, whereas affective cultures elevate emotional expression over neutrality (Trompenaars and Hampden-Turner, 1997). Verbal or non-verbal communication occurs differently in affective and neutral cultures. Having grown up in the 'emotional cultures' where a full range of emotion from laughter to anger is both human and acceptable, French respondents rated the *Display of Feelings* dimension highly. People from this society are not ashamed to exhibit their feelings, for example by intentionally touching others, by displaying emotions publicly or by discussing sensitive issues. They have open relationships with others and speak freely about their feelings and personal experiences. They like to *Compliment* others whom they believe are deserving. By contrast, people in Eastern societies are indirect in behaviour, strive to maintain harmony in human relations and follow the ethic of not questioning, disagreeing or hurting the feelings of others. They avoid paying compliments because compliments can cause harm. The 'neutral culture' urges Asian people to be objective, detached and rational, since emotional expression clouds reason. The Vietnamese attach greater importance to regulating social life than is the case in most Western countries. Central concepts include *Kinh trong, Le phep, Nha nhan*. This refers to being refined, polite, and perfect in behaviour. The person should be restrained and calm. People should keep a steady state which concerns the correct manner of doing things by being calm and steady, avoiding extremes of emotional expressions and of activity.

In terms of *Form of Greeting*, French usually shake hands with others in workplaces or in public. The French are usually very relaxed and do not adhere to strict or explicit codes of behaviour or ceremony (Harris and Moran, 1996). Similarly, having become accustomed to the French and

American cultures, Vietnamese respondents also shake hands, particularly in the workplace. Moreover, since the Vietnamese have recently become familiarised dealing with Western businessmen, particularly during the period economic reforms, they politely accept handshaking.

Eastern and Western societies hold different views towards privacy. These disparities arise because people have grown up in different cultural backgrounds. French respondents attached highest importance to *Respect Privacy*. French emphasize specific issues in a relationship, addressing personal background later (or not at all), limiting engagement to specific areas of life. In these societies, people separate work sharply from other areas of life. People focus on "business only" and friendships may perhaps form later. They want to get straight to the point. Getting to know a person and their various relationships is possible, but not essential. The value of individuality leads the French penchants for self-improvement and for protection of individual privacy (Samovar and Porter, 1991). The French pattern of relationships is the reverse of Asian cultures, where solitude is often perceived positively, and other people's privacy and individual activities are respected.

In a 'diffuse' and 'collectivist' culture such as Vietnam , personal relationships are inclusive. Intensely personal, exclusive friendships are not encouraged. Interpersonal relations develop within the context of village and community. Society is very much oriented around group activity. In these societies, people attach closer to each other and tend to show considerable interest in the private affairs of others. At a certain point, this nice and considerate attribute can become prying and interfering and make others uncomfortable and annoyed. *Respect Privacy* is not considered as a critical matter in Vietnamese society, since people engage simultaneously in multiple arenas of life. They usually begin with personal background, eventually focusing on specific issues in view of the rapport established. They ask about family, children and politics prior to discuss business. They want to get to know a person first. A person's character is revealed through friendship and this forms the basis for business transactions. Establishing a good relationship or "Guanxi" is essential to doing business in Vietnam.

The French society marvel at modem comforts which include air-conditioning, microwave ovens, and swift and convenient transportation. Material comforts signify the attainment of success (Engel et al., 1995; Schiffman and Kanuk, 1991). With sufficient determination, responsibility and initiative, an independent individual may control his or her own destiny.

French value personal identity and, to an extent, self-worth can be measured by achievements and success (Engel et al., 1995; Harris and Moran, 1996, Barsoux and Laurence, 1990a). The historical roots of the achievement value can be traced back to religious beliefs (i.e., the Protestant and Catholic Work Ethic). Hard work is viewed as wholesome, spiritually rewarding, and as an appropriate end in itself. These values generally serve as a moral and social justification for conspicuous consumption of goods and services, and are epitomized in popular expressions such as, 'You deserve it,' 'You owe it to yourself,' and 'You worked hard for it' (Schiffman and Kanuk, 1991).

The French emphasize the importance of 'being themselves,' and celebrate their individuality and uniqueness. They idealize the self-made person who is self-reliant, self-interested, self-confident, self-fulfilling, and possesses self-esteem. By contrast, the strict principles of social stratification in Vietnamese society emphasises communal consensus, mutual togetherness, and sociability and implies doing things in the company of others, never being alone and sacrificing individual needs and wants for the sake of the group. French reject the idea of dependency and believe that individuals should be self-reliant. As a result, French focus on being independent, are orientated towards individuals and attach great importance to *Individual Needs*. This was confirmed in Hofstede's (1984) study which reported that in countries with low Power Distance and Low Uncertainty Avoidance, people are more likely to view the ideal organisation as one in which power is spread throughout the organisation. This defused power is based on personality and competence, where activities are sufficiently loosely structured to allow for change and individual initiative, much like a market model with implicitly structured activities.

TOURISTS-HOSTS INTERCULTURAL INTERACTIONS

The research has indicated a divergence of interactions between French tourists and Vietnamese hosts. The dissimilarity exhibited within the French tourists is not only attributable to the impact of particular cultural values and rules of behaviour of each nationality, but may be attributable to the travel motivation of each market. For instance, the French Tourist travelled to Vietnam primarily for pleasure purposes. These tourists may wish to interact more with Vietnamese hosts and the Vietnamese local people as part of their search for the "exotic." The perceived exoticism is also magnified by the long haul travel that is required to reach Vietnam. Due to strong previous attachment to Vietnam, French respondents also enjoyed interacting with their hosts as a way of recalling past memories or finding out about how things have changed since the Vietnam War.

Along with the various motivations noted above, different cultural values and rules of behaviour are major explanations for the different level of interaction exhibited across the French tourists with their Vietnamese hosts during the service encounters. According to Parsons (1973), the Universalism-Particularism dimension differentiates cultures based on how people describe others, and the rules they use for this purpose. In universalistic cultures such as France, people interact and communicate with strangers in the same way regardless of social situations and circumstances, while in particularistic cultures such as Vietnam, the interaction and communication patterns with strangers vary according to the situation. The Instrumental-Expressive Orientation dimension also differentiates between cultures depending on the nature of the goals that people are seeking in their social interactions (Parson, 1973).

According to Argyle (1967), when the participants in service encounters are culturally the same or similar, they may reject each other if they do not conform to each other's cultural patterns of interaction and expected standards. Many French would like to interact and communicate more with their Vietnamese hosts. This is unsurprising, considering that they are by nature less formal, more talkative, entertaining and humorous. According to the French guests, more often than not, the emphasis for both hosts and guests is on the self and its accompanying distinct and unique qualities such as being easy-going, fun loving and outgoing during the service encounter. Given the appreciation of talkativeness amongst Western guests during service encounters, it is not surprising that other researchers have found that "extended interpersonal encounters elicit interpersonal exchange and displays of positive and esteem-enhancing emotions" (Price, Amould and Tierney, 1995, p. 85). French guests and Vietnamese hosts allowed more time for self-revelation during these extended service encounters.

On the basis of the differences noted above, Vietnamese hosts need to be better prepared to cater for the French tourist market. Communication services could be considered as one of many important elements for assessing the cross-cultural perceptions and satisfaction of tourists. This implies that Vietnamese hosts will need to improve their communication skills in order to achieve a higher level of social or professional interaction with guests during the provision of service. This could help to establish a good understanding of host expectations, perceptions and satisfaction towards their products and services.

COMPARATIVE PERCEPTION AND SATISFACTION WITH SERVICE QUALITY

It is evident from the results presented in Tables 5 and 6 that hosts and guests from different cultural backgrounds evaluate host service attributes and performance differently either consciously or unconsciously. When the relevant service indicators were examined in relation to the Vietnamese service providers and the, French guests, distinct dimensions were identified. These dimensions were perceived by Vietnamese hosts and the French sample to be components of the perceived importance and satisfaction with actual experience towards the service attributes and performance. However, each dimension differed in terms of its measures, and level of influence on each of the two constructs cross-culturally. The dimensions and measures that were generated for each of the two cultural groups provided valuable insight into what service providers need to do to in order to influence positively tourists perceptions and satisfaction with service attributes and performance in each of the given cultures. Each of these dimensions, in relation to each of the two cultural groups, is discussed below.

For the Vietnamese culture, the hosts were very satisfied with the attribute 'Attentiveness' which was placed as the first dimension in both importance and actual experience. In addition, there are wider ranges of service attributes that gained from the actual experiences which are outweighed the Vietnamese hosts' level of perceived importance. For instance, the 'Intercultural Competence' which was placed as fifth dimension in the degree of importance was found to be the second dimension in the degree of actual experience. The hosts were also satisfied with dimension 'Treat Tourists as Guests' which was placed as the fifth dimension in their actual experience even though they had not considered this dimension as being of high importance. Conversely, they were not satisfied with the attributes 'Responsiveness & Confidence' and 'Trustworthiness' which were placed some distance away as third and sixth in order in their actual experience, even though these dimensions had been rated as of high importance.

French tourists attached high importance to a number of service attributes. These consist of 'Courtesy & Friendliness', 'Verbal Communication', 'Helpful & Trustworthiness', 'Accurate Information' followed by 'Intercultural Competence & Personalized Attention' and 'Capability'. In their actual experience, they were satisfied with the attributes 'Courtesy & Friendliness' and 'Helpful & Trustworthiness'. Moreover, French tourists were very satisfied with the dimension 'Communicative' as it outweighed their level of importance, even though they had not considered this dimension as important.

It was notable that French tourists rated 'Intercultural Competence & Personalized Attention' as the fifth most important dimension, but were not satisfied with it in their actual experience. In addition, negative results have also showed on a numbers of dimensions including 'Personalized Attention', 'Verbal Communication', 'Accurate Information', 'Capability', 'Confident & Anticipate' and 'Responsiveness' in the actual experience, indicating that

French tourists were not as satisfied with these dimensions as they might have expected.

In summary, the outcomes presented in Tables 5 and 6 support the view that tourist satisfaction with service was obtained as a comparison of perceived importance and actual experience of host service attributes and performance. There are considerable differences in the dimensions of importance and satisfaction with actual experiences between the two cultural groups studied in terms of service. The outcomes indicated that there is a greater disparity in terms of importance and satisfaction between the Vietnamese hosts and French tourists. Clearly the distinction of cultural values and rules of behaviour of respondents from different cultural backgrounds have an important role to play in the levels of satisfaction with tourism services, and this relationship is an indirect way, related to the different levels of importance placed by different cultural groups on Vietnamese host service attributes and performance. These different levels of importance are associated with different attitudes to satisfaction with actual tourism services consumed or offered.

According to Kluckhohn and Strodtbeck's (1961) and Stewart (1971), the differentiated activity orientations are evident between cultures on the basis of how people view human activities and how they express themselves through activities such as "doing/being/becoming cultures." This orientation determines the pace of people's lives and influences the relationship between work and play. Western societies such as France are 'doing' and 'action' oriented cultures. In these societies, people engage in spontaneous activities, indulge in pleasures and reveal their spontaneity as an expression of their human personality. They emphasize activity, task completion, goals achievements, getting things done, and competition. Activities are tangible and can be externally measured. People decisions are most likely to be economically driven and task oriented. The activity orientation dimension has a significant influence on their interpersonal relations. They seek change and want to control their lives. They are governed by time schedules and appointments, and are characterized by a fast pace of life. For this reason, work is separated from play because employees are supervised and controlled. In these societies, challenge solves a problem when a difficulty occurs; and the interpersonal interaction and communication are characterized by accomplishing specific tasks and solving problems adequately.

By contrast, Vietnamese society belongs to "Being" cultures. In this society, people focus on non-action and believe that all events are determined by fate. They are concerned with spiritual life more than a material one. They are concerned with who people are, and not what they have. They accentuate individual obligations to society as their decisions are most likely to be emotional and people oriented. They also lay emphasis on passivity, defensiveness and striving for social harmony in interpersonal relations at the expense of efficiency. Belonging to a "Being" activity oriented culture, Vietnamese people have different approaches to work and leisure. They are characterized by a slower and more relaxed pace of life, and they consider the process to be more important than the final result. According to them, work is a means to an end and there is no clear distinction between work and play and they accept the difficulty rather than challenge and eliminate it. Consequently, in the work place, employees can mix together and socialize and social interactions are characterized by being together.

Based on their cultural orientation, French respondents expected precise and accurate information. In contrast, Vietnamese hosts are less concerned with providing such details and are not worried if problems are not resolved immediately even though that they have rated highly the dimensions *Knowledgeable and Communicative* as well as *Capable and Trustworthy*. Coming from an egalitarian society with a strong focus on logic and science, French guests are more direct and open. While the straightforwardness is absolute for them, truthfulness is relative for the Vietnamese. For that reason, *Accurate Information* are French tourists' expectations that their hosts have to be informative and adequate in the information provided, be trustworthy, approachable and be able to solve problems quickly. To be *Knowledgeable* is an important attribute for the Vietnamese service providers, particularly in the case of tour guides as this job requires them to answer any tourists' enquiries accurately, because there are some Vietnamese hosts who may be unfamiliar or even impassive about their own history. For many French tourists, and in particular, for the War Veterans and their families, the long and complex history of Vietnam, with so many historical sites and war relics, is considered as highlight of the country. However, the result from the interviews and observations indicated that some respondents were disappointed by the one-sided approach to history shown in some museums. A balanced and sensitive interpretation from the Vietnamese hosts is therefore important to provide the appropriate experience for visitors who arrive with diverse needs and expectations.

Vietnamese hosts attached high importance to the dimension *Communicative*. This attribute has also been confirmed by French respondents who reported that their tour guides were of high quality. The majority of Vietnamese tour guides were commended for their mastery of English. However, this is not the case for French respondents who reported in the opposite way. Tour guides were regarded as playing an important role in transmitting local knowledge and in helping tourists to avoid uncomfortable contacts with Vietnam hosts. Good tour guides have the ability to interpret and to give visitors useful information about tourist sites. This was found to have an influence on their satisfaction with their learning, especially with their cross-cultural experiences. However, a number of French respondents did comment that some tour guides seemed to be reciting scripts or trying to be politically correct and lacked their "personal touch." Such actions created doubt and distrust in the minds of visitors among the visitors. According to Schmidt, a good guide can provide tourists with authentic tourist experiences and psychological satisfaction. In addition to the language skills and the knowledge required for their jobs, tour guides should have good cross-cultural understanding and have enthusiastic and pleasant personalities (1979). Regular training should be given to tour guides to enhance their knowledge and improve their interpretation skills and help them become familiar with the peculiarities of their customers, in this case, French guests.

In most Asian societies, a complex system of grading based on age, occupation, and positions within Vietnamese society requires correct behaviour and respect to be shown towards all of higher social standing within the vertical social hierarchy. Within this highly complex system of deference to hierarchical authority, customary laws and standards specify how each community member should react to others, the nature and forms of obligations within and outside kin groups. Each age group must, for example, be addressed by the correct terminology and language. In contrast, French respondents may not readily understand certain principles of social stratification and age grading within Vietnamese society. Being accustomed to greater egalitarianism, French people are more casual and their behaviour depends less on social position and age. Consequently, their expectations in terms of the *Courtesy and Friendliness* dimension are to be treated by their Vietnamese hosts with an approachable, friendly, considerate, confident and polite manner as an expression of respect and kindness rather than an expression of formal etiquette. This is could be explained in that to most Western societies, such as France, individuality and freedom seem to be positively balanced and this is evident in the way that people communicate with others, for instance with their peers, family and friends. French culture has a strong sense of history which is rooted in democratic ideals, such as 'freedom of speech,' 'freedom of press,' and 'freedom of worship.' (Schiffman and Kanuk, 1991). As an outgrowth of these democratic beliefs in freedom, the French has a strong preference for freedom of expression - the desire to be oneself and to be responsible to oneself (Schiffman and Kanuk, 1991). In the service encounters, more often than not, the emphasis, for both hosts and guests, is on the self and its accompanying distinct and unique qualities such as being easy-going, fun loving and outgoing. In this culture, a more open and frank style of communication is esteemed. For these reasons the French people are generally described as loud, flamboyant, confident, outgoing and effervescent.

The Vietnamese concept of politeness is very different from what prevails in Western societies. It has been suggested that the Vietnamese concept of politeness encompasses four main qualities: respectfulness, which includes concern for others' face and status; modesty and humbleness which include self-effacing or abasing behaviour; attitudinal warmth including a display of kindness, consideration and hospitality; and refinement which includes display of manners and civil behaviour, and being cultured. These four qualities are considered extremely important in the

Vietnamese culture and custom. In Vietnamese society, people generally adhere to the traditional ways of behaviour and formal etiquette as well as to other person's status. Not surprisingly, the Vietnamese today still prefer to address someone in more hierarchical-type terms such as 'uncle' and auntie,' or 'sir' or 'madam', rather than by their surname. Consequently, Vietnamese hosts rated high importance for the dimension *Smart and Polite* since they placed great emphasis on the tangible cues associated with their service such as physical appearance (e.g. neatly dressed), and behaving politely towards their visitors. According to the Vietnamese customs, foreign visitors need to be treated with an extraordinary hospitality, respect, courtesy and warmth manner. For this reason, it is not surprising that Vietnamese hosts rated highly on the dimensions *Attentiveness* as well as *Treat Tourists as Guests*. Based on Vietnamese hosts perspectives, visitors' needs should be fulfilled with intensive attention and care in order to make a guest happy and satisfied. In contrast, this attitude many not be appropriate with French visitors as they believe that individuals know best how to satisfy their needs.

In Vietnamese society, individuals strictly observe the accepted behavior which is governed by a hierarchy of social structures, roles and relationships (Yates and Lee, 1996). However, French behaviour depends less on social position and age, and their clothing style is more casual because French society is more egalitarian. Feelings of duty and obligation may be less prominent and rules of formal etiquette are limited in this society. French people conduct their lives according to other principles and manners, and therefore such etiquette of social behaviour plays a relatively minor role in their lives. The rules of formal behaviour are not generally acceptable because prevailing behaviour is more casual and depends less on social position and age. People therefore place less importance on physical appearance and become conversant with themselves on informal behaviour.

With reference to human time orientation, the efficiency and timeliness of service performance are considered as essentially two important service aspects. The speed and efficiency of a transaction have been addressed by many researchers as important elements in evaluating service quality (Solomon et al., 1985; Taylor, 1994). Efficiency has also been presented as a dimension that consumers perceive as a trade off for personalization (Sutton and Rafaeli, 1988). Although most tourists would appreciate efficient and timely service, they may have different tolerance levels or thresholds. This is because French society perceive time as monochrome, sequential, absolute and prompt (Redding, 1980). In other words, Western societies tend to be more sensitive to time, procedure, schedule and deadlines (Kirkbride et al., 1991).

Even though French have viewed time as synchronization and often see life "as-a-dance" (Hampden-Turner and Trompenaars, 1993) they also attached high importance on *Responsive & Punctual Service*. By contrast, Vietnamese synchronize all efforts to get things done and try to do many things simultaneously, treat time as elastic and use the past to advance the future. For this reason, time is stretchable and, as a result, time commitments do not have to be kept in the case of the Vietnamese hosts. People are usually flexible about time, appointments and the provision of service. Except for the services offered in some 4 and 5 star accommodations or resorts, most activities, including the service provided by other sectors like restaurants and retail shops, may occur over an extended time period, continuing for at least twice as long as the corresponding Western activity. This may be part of the concept in Vietnamese culture that being in a hurry and looking for quick solutions to problems is an indication of impatience. The Vietnamese style of perceiving time is more flexible and in a more relaxed manner rather than the French idea of 'a time and place for everything'. For French guests' perspective on *Prompt & Punctual Service* involved the host ability to solve problems quickly and to answer all questions to tourists. As a result, the focus on punctuality and efficiency of service provision in responding to clients' needs in a satisfactory manner is very critical for the Vietnamese hosts.

According to Parson's pattern variables (1951/1953), in the highly individualistic and vertical cultures with achievement orientation such as France, people and objects are judged on the basis of the performance and measurable results, and others' behaviour is predicted on the basis of their efforts and occupational status and achievements. People categorize others according to specific details such as facts, tasks and numbers and respond to a particular aspect of a person or an object only, for instance, the role or responsibility assigned. In these societies, people express self-discipline and their behaviour and decisions are guided by cognitive information and facts. In France, people believe that each individual is unique, special and completely different from all other individuals, thus the interests and the needs of the individual are paramount. They believe that individual satisfaction comes from personal achievement. They are successful in controlling the natural environment; they attempt to dominate space as well. They think that human nature can be changed and education is an important element in improving human nature. Importance is placed on learning to be an individual, independent, self-motivated and achievement oriented. They are very competitive and assertive and they always rank, grade, classify and evaluate to know if they are the best.

Due to the above reasons, there are also many other dimensions that French respondents have expected from their Vietnamese hosts including *Capability, Trustworthy, Confident & Anticipate, Accurate Information* and especially to be *Knowledgeable.* All these dimensions can be summarised as the vital elements of *Capability* that the service providers must have in order to be a professional employee. This also entails the required knowledge, skill and training that hosts must perform efficiently in order to achieve effective and reliable service delivery. One important aspect of *Capability* and competence is hosts' attitude towards work. Hosts' attitude can have a huge impact on guest perceptions of service quality. Service providers must be aware that their attitude towards work can easily be communicated to guests through their behaviour and verbal cues. This is especially important in a cross-cultural context, where service providers should be trained to be culturally sensitive and be aware of cultural differences, as well as similarities. In light of this, being equipped with the necessary knowledge, skills, and possessing a positive attitude, will allow hosts to better understand their guests and where possible, accommodate their requirements.

The production and distribution of services in international tourism involves a substantial amount of cross-cultural experience on the part of both tourists and service providers. Despite the friendliness and courteousness of

Vietnamese service providers, in general staff appear to lack *Intercultural Competence* and experience in dealing with Western customers. Regardless of the fact there is a high proportion of Vietnamese tour guides who were commended for their mastery of English, the ability to speak other languages including French and even English amongst the Vietnamese service providers in general still appeared unsatisfactory. To be fluent and have a command of the foreign languages is very critical particularly when catering to the international tourist markets. The French guests have placed high importance on *Communicative* and *Verbal Communication* but they were not satisfied with these dimensions. French people are very conservative, and because of the national pride on their mother tongue, they prefer to speak to other people in their own language. For that reason, *Capability* from French tourists' perspectives is that hosts are able to perform services and to keep them informed in their language.

Vietnamese hosts and the French tourists have recognised *Intercultural Competence* as an important dimension. Although they rank this factor differently, similarities are evident between the two groups, possibly attributable to the strong prominence of the service interaction in the cross-cultural setting. Furthermore, it is noticeable that the dimension on *Intercultural Competence* is linked to *Verbal communication* in the case of Vietnamese hosts, whereas it is linked to to *Personalized Attention* in the case of French guests. The distinction of the two groups could be explained by French tourists holding different views on perceptions and satisfactions towards hosts' service attributes and performance.

From a collectivist and hierarchal culture such as Vietnam, the importance of a person depends on his or her social position, age and gender. Social rank determines the manner in which people will be perceived and treated. Social respect is gained through status and age, which are symbols of experience and wisdom. Respect and deference is given to authority and high hierarchy positions. Nonetheless, in an individualistic society such as France, people know best what their needs are and how these needs can be satisfied. The need to think and behave like individuals and to preserve one's privacy cannot be understood by the Vietnamese society because the concept of privacy does not even exist in this culture. In the French society, social recognition is gained through hard work and achievement. Being self-reliant, self-interested and having self-esteem, French guests expected that their hosts would treat them with friendly and individualized attention. They expected a customized service and personal attention that could enable them to express their own personality.

Despite the fact that French tourists enjoyed the distinctive Vietnamese culture, they were unhappy about the inability of Vietnamese to communicate with them in some retail shops or restaurants. As members of high uncertainty avoidance cultures, the Vietnamese worry about being exposed to language difficulties when serving foreign tourists. Additionally, the cultural imperative, in Vietnamese society, to care for foreign visitors results in a sense of responsibility to give constant attention to foreigners. Beyond a certain point, this may become annoying for French visitors. For instance, the sellers follow every step of tourists in their shops as a signal of personalized attention to guests, but it may appear as an untrustworthy and pushy manner from the tourist perspective. Enquiries about peoples' age and earnings are acceptable in Vietnamese society and are viewed as signs of thoughtfulness. However, they are considered to be impolite in the individualist French culture where personal privacy is respected. As a result, the inherent need to care about foreign visitors in Vietnamese culture results in a national responsibility for giving constant attention to and helping foreigners to cope with the different customs, to a degree that may become annoying for French tourists as they might feel uncomfortable when someone else decides about fulfilling their needs. Consequently, it is imperative that knowledge of French guests' cultures and languages, as well as the *Intercultural Competence* amongst the Vietnamese hosts is essential when responding to guest standards of behaviour and needs.

Principal Components Analysis (PCA) was used to determine whether or not the basic factoral structure of cultural differences between the Vietnamese hosts and the French tourists existed and was significant. The factorial analysis was applied only for variables which differed between the Vietnamese and French samples. The analysis identified a clear number of dimensions (groupings) of cultural values, rules of behaviour, perceptions of and satisfaction with service components that differed between the

Vietnamese hosts and the the French tourists. The analysis also identified a clear number of cultural dimensions in French language group that differed from the Vietnamese sample.

The above results explain each cultural and behavioural dimension identified by the Principal Components Analysis (PCA). The outcomes have also provided additional insights into these differences by explaining how and why the French guests differed from the Vietnamese hosts on each cultural dimension and behavioural dimension.

Chapter 6

CONCLUSION AND IMPLICATIONS

The differences that have been identified between Vietnamese hosts and French guests are broadly consistent with the literature, and substantiate some popularly held views about cultural differences between the Asian and Western worlds. The research reported in this chapter has clearly identified that Vietnamese hosts and French tourists have different cultural values and rules of behaviour. The variables that loaded significantly on the above dimensions are the key cultural values and rules of behaviour determinants of the service encounter interactions and the mutual perceptions between hosts and guests.

There has been little empirical evidence to date to demonstrate such dissimilarities in the Vietnam context. The present study has contributed to the body of knowledge concerning differences between culture and rules of social interactions in the tourism context. It has been commonly believed that perceptions and satisfaction with tourism products and services are subjective and influenced by aspects other than culture and rules. The present study has indicated that this is not always the case. Cultural factors in conjunction with the rules of behaviour should underpin the development of any new hypothesis or theory about the intercultural interactions accompanied by the perceptions and satisfaction with products and services received, and how tourists respond to current offerings.

The current research supports the notion that different cultural values and rules of behaviour between hosts and guests form an important assessment index for host-guest service encounter interactions, along with the inter-perceptions (or mutual-perceptions) and satisfaction with products and services offered and consumed in the cross-cultural tourism context. The study findings should provide tourism service providers and managers with useful guidelines to enhance the cross-cultural understanding of tourist service evaluations. Based on reliable theory and practice in the context of social processes, these recommendations should enhance competitive advantage for businesses and destinations when targeting a multicultural clientele.

The present research has recommended that the tourism industry should re-evaluate its marketing practices, focus on the different cultural values and rules of behaviour, and examine the impact of such differences on host-guest service encounter interactions together with the mutual-perceptions in terms of products and services. Besides cultural values, an important implication for tourism marketers is that rules of behaviour, perceptions and satisfaction are fundamental and useful constructs for market segmentation and promotion. The similarities and differences of cultural values and rules of behaviour between the Vietnamese hosts and French tourists should be considered when developing tourism marketing strategies, because the by-products of these different social interactions are perceptions which determine tourist behaviour and decision-making.

In the case of developing countries such as Vietnam, cultural training may assist service providers to understand their own culture as well as the culture and rules of behaviour governing tourist generating countries and to appreciate these differences. Such awareness of culture and rules of social interaction will enhance host understandings of international tourists emanating from different cultural backgrounds. This should minimise any negative perceptions and dissatisfaction resulting from misunderstandings of tourist psychological needs and experiences.

When marketing internationally, tourism marketers would be well advised to consider the different cultural values and rules of behaviour, which apply to service encounters. Marketers can directly influence perceptions and satisfaction through addressing prevailing cultural values for instance 'Esteem & Personal Contentment', 'Salvation', 'Competence', 'Sense of Accomplishment', 'Sense of Self', 'A World of Beauty', 'Hedonistic Values (for European guests in general and French Tourists in particular). Furthermore, marketers should deliberate different rules of behaviour upon hosts-guests intercultural interactions including 'Interest in Others', 'Respect for Privacy' 'Clarity of Expression', 'Display of Feelings', 'Compliment', 'Individual Needs', 'Form of Greeting' and 'Communication Behaviour' (especially in the case of the French guests and the European tourists at large).

Since these influences vary from market to market, it would be useful to replicate the present study using samples drawn from other Asian and Western source markets. It is becoming increasingly important for tourism marketers to understand cultural values and rules of behaviour in order to create service quality perceptions and service encounter interactions, which may in turn lead to overall satisfaction. The present study has highlighted the need to explore the perspectives of international tourists from diverse cultural backgrounds towards destinations, and develop culture-oriented marketing strategies appropriate for these markets.

As destinations experience increases in tourist arrivals from diverse cultural backgrounds, there is increasing pressure on service providers to cater for the needs of tourists who bring different expectations. To satisfy the multicultural needs and wants of international tourists, to enhance repeat visitation and stimulate favourable word of mouth recommendation, service providers will need to pay greater attention to cultural differences. It is hoped that the current research will provide marketing insights for international service providers across Southeast Asian where the historical and cultural settings are similar to Vietnam.

A number of study limitations should be acknowledged. Vietnam is in its early stages of development as a holiday destination. To keep the present research within manageable bounds of field word, the scope of this study was confined to few major cities within Vietnam. The various cities have hotels, restaurants, airports and retail outlets, but the 'climate' of these settings is distinct and may have impacted on guest service evaluations.

In part because of the ambitious coverage of two nationalities, the research for each group drew upon on a relatively small sample, especially in the case of the French (180). The sample sizes for the Vietnamese (205) population was above 200 and conform more closely with established norms. The exploratory nature of the study precluded the use of large samples. Though stronger representativeness might have provided greater generalizability, the study findings have provided a statistically sound basis for purposes of generalization. The conduct of the research in various settings including airports, hotels, restaurants, shops and tourist attractions limited the exercise of control by the researcher. The interviewees were pleasure travellers who valued their time and business travellers were particularly time constrained.

As the extent of influences varies from market to market, it would be useful to replicate the study on samples of various Asian and Western nationalities to determine these influences and form a basis for appropriate promotional strategies. An important implication of this finding is that it becomes increasingly important for tourism marketers to have knowledge on the dissimilarity of cultural values and rules of behaviour to create service quality perceptions as well as service encounters interactions. This may in turn lead directly to overall satisfaction. The significance of the study lies in highlighting the need for further exploration the differences in cultural values and rules of behaviour amongst international tourists from diverse cultures and developing culture-oriented marketing strategies.

REFERENCES

Ackoff, R. L. & Emery, F. E. (1972). *On Purposeful Systems.* London: Tavistock Publications.

Adler, F. (1956). The Value Concept in Sociology. *American Journal of Sociology,* 62, 272-279.

Adler, N. J. (1997). *International Dimensions of Organisational Behaviour,* Cincinnati, Ohio: International Thomson Publishing.

Albert, E. M. (1956). The Classification of Values: a Method and Illustration. *American Anthropologist,* 58, 221-248.

Albert. E. M. (1968). Value Systems. In Sills, D. L. (Ed.) *International Encyclopedia of the Social Sciences 1.* New York: MacMillan, 287-291.

Albert. R. D. & Triandis, H. C. (1979). Cross-Cultural Training: A Theoretical Framework and Some Observations. In Trueba. H. and Bamett-Mizrahi. C. (Eds.) *Bilingual Multicultural Education and the Professional: From Theory to Practice.* Rowley, Mass: Newbury House.

Allport, G. W., Vemon, P. E. & Lindzey, G. (1951/1960). *A Study of Values: A Scale for Measuring the Dominant Interest in Personality.* Boston, Mass.: Houghton Mifflin Co.

Allport, G. W. (1961). *Pattern and Growth in Personality.* New York: Holt, Rinehart and Winston.

Anderson, E. & Weitz, B. A. (1990). Determinants of Continuity in Conventional Industrial Channel Dyads. *Marketing Science,* 8(Fall), 310-323.

Argyle, M. (1967/1978/1990). *The Psychology of Interpersonal Behaviour.* Penguin Harmondsworth: Books Ltd.

Argyle, M. (1986). Rules for Social Relationship in Four Cultures. *Australian Journal of Psychology,* 38 (3), 309-318.

Argyle, M. & Henderson, M. (1985a). *The Anatomy of Relationships: And the Rules and Skills Needed to Manage Them Successfully.* London: Heinemann.

Argyle, M. & Henderson, M. (1985b). The Rules of Relationships. In Duck, S. and Perlman, D. (eds.) *Understanding Personal Relationships.* London and Beverly Hills, CA: Sage Publications: 63-84.

Armstrong, R., Mok, C., Go, F. & Chan, A. (1997). The Importance of Cross-Cultural Expectations in the Measurement of Service Quality Perceptions in the Hotel Industry. *International Journal of Hospitality Management,* 16(2), 181-190.

Ashworth, G. & Goodall, B. (1988). Tourist Images: Marketing Considerations. In: Goodall, B. and Ashworth, G., Editors, 1988. *Marketing in the Tourism Industry, the Promotion of Destination Regions.* Routledge, London, 213-238.

Assael, H. (1992). *Consumer Behaviour and Marketing Action* (4th Ed). Boston: PWS-Kent.

Atkinson, T. & Murray, M. (1979). *Values, Domains and the Perceived Quality of Life.* Paper presented at the Annual Meeting of the American Psychological Association, New York.

Babakus, E. & Boiler, G. W. (1992). An Empirical Assessment of the SERVQUAL Scale. *Journal of Business Research,* 24, 253-268.

Baier, K. (1969). What is Value? An Analysis of the Concept. In Baier, K. and Rescher. N. (Eds.) *Values and the Future. The Impact of Technological Change on American Values.* New York: The Free Press.

Bailey, J. (1991). *Managing Organizational Behaviour.* 2nd Ed. Brisbane: John Wiley and Sons.

Bamlund, D. C. & Yoshioka. M. (1990). Apologies: Japanese and American Styles. *International Journal of Intercultural Relations,* 14(2), 193-206.

Barsoux, J. L. & Lawrence, P. (1990a). *Management in France.* Cassell Educational Limited, London.

Barth, F. (1966). *Models of Social Organization.* Occasional Paper No. 23, Royal Anthropological Institute, Norway.

Barton. A. (1969). Measuring the Values of Individuals. In Dean. D. G. (Ed.) *Dynamic Social Psychology.* New York: Random House.

Befu, H. (1971). *Japan: An Anthropological Introduction.* New York: Harper and Row.

Bharadwaj, L. & Wilkening, E. A. (1977). The Prediction of Perceived Well-Being. *Social Indicators Research,* 4, 421-439.

Bennett, P. D. & Kassarjian, H. J. (1972). *Consumer Behaviour.* Englewood Cliffs, New York: Prentice Hall.

Berthon, P. (1993). Psychological Type and Corporate Culture: Relationships and Dynamics. *Omega,* 21, 329-344.

Berry, J. W., Poortinga, Y. H., Segall, M. H. & Dasen, P. R. (1992). *Cross-cultural Psychology: Research and Applications*, Cambridge University Press, Cambridge/New York.

Black, J. S. & Mendenhall, M. (1989). A Practical but Theory-Based Framework for Selecting Cross-Cultural Training Methods. *Human Resource Management,* 28(4), 511-539.

Bochner, S. (1982). Cultures in Contact: Studies in Cross-Cultural Interaction. Oxford; New York: Pergamon Press.

Boissevain, J. & Inglott, P. (1979). Tourism in Malta in De Kadt, E. (Ed.), *Tourism: Passport to Development?* Oxford University Press, Oxford.

Bond, M. H. & Hwang, K. (1986). The Social Psychology of Chinese People, in *The Psychology of the Chinese People,* Ed. M.H. Bond, Hong Kong: Oxford University Press, 213-266.

Bond, M. H. & Chinese Culture Connection (1987). Chinese Values and the Search for Culture-Fee Dimensions of Culture. *Journal of Cross-Cultural Psychology,* 18(2), 143-174.

Brewer, J. (1978). Tourism Business and Ethnic Categories in a Mexican Town, in Smith, V. (Ed.) *Tourism and Behaviour*. College of William and Mary, Williamsburg, VA.

Brislin, R. W, Lonner, W. J. & Thorndike, R. M. (1973). *Cross-Cultural Research Methods*, New York, J. Wiley.

Brown, T. J. Churchill, J. A. & Peter, J. P. (1993). Improving the Measurement of Service Quality. *Journal Retailing*, 69(1), 127-139.

Bruner, G. C. & Hensel, P. J. (1993). Multi-Item Scale Usage in Marketing Journals 1980- 1989, *Journal of the Academy of Marketing Science*, 21(Fall), 339-344.

Callan, R. J. (1997). *An Attributional Approach to Hotel Selection. Part 1: The Managers' Perception.* Progress in Tourism and Hospitality Research, 3, 333-349.

Camilleri, C. (1985). La Psychologie Culturelle [Cultural Psychology]. *Psychologic Francaise*, 30, 147-151.

Campbell. D. T. (1963). Social Attitudes and Other Acquired Behavioral Dispositions. In Koch, S. (ed.) *Psychology: A Study of Science. Investigations of Man as Socius, Their Place in Psychology.* In: The Social Sciences 6. New York: McGraw-Hill.

Carman, J. M. (1990). Consumer Perceptions of Service Quality: An Assessment of the SERVQUAL Dimensions. *Journal of Retailing*, 66, 33-55.

Catalone, R. J., Di Benedetto, C. A. & Bojanic, D. C. (1989). Multiple Multinational Tourism Positioning Using Correspondence Analysis. *Journal of Travel Research*, 28(2), 25-32.

Cattell, R. B. (1978). *The Scientific Use of Factor Analysis in Behavioural and Life Sciences*, New York: Plenum Press.

Cattell, R. B. (1953). A Quantitative Analysis of the Changes in the Culture Pattern of Great Britain, 1837-1937, by P-Technique. *Acta Psychologica*, 9, 99-121.

Catton, W. R. (1959). A Theory of Value. *American Sociological Review*, 24, 310-317.

Chamberlain, K. (1985). Value Dimensions, Cultural Differences and The Prediction of Perceived Quality of Life. *Social Indicators Research,* 17, 345-401.

Cheng, T. C. (1988). A Theoretical Analysis and Empirical Research on the Psychology of Face [in Chinese], in *Chinese Psychology,* Ed. K. S. Yang, Taipei: Laureate Book Co., 155-237.

Chinese Culture Connection (1987). Chinese Values and The Search for Culture-Free Dimensions of Culture. *Cross-Cultural Psychology,* 18(2), 143-164.

Chung, C. Y. (1969). Differences of the Ego as Demanded in Psychotherapy, in the East and West. *Psychologia,* 12, 55-58.

Churchill, G. A. (1979). A Paradigm for Developing Better Measures of Marketing Constructs. *Journal of Marketing Research,* 16(February), 64-73.

Clawson, G. & Vinson, D. (1978). Human Values: A Historical and Interdisciplinary Analysis. In K. Hunt (Ed.). *Advances in Consumer Research,* 5, 396-402. Michigan: Ann Arbor.

Cooper, M., and Hanson, J. (1997). Where there are no tourists…yet. A Visit to the Slum Brothels in Ho Chi Minh city, Vietnam. In M. Oppermann (Ed.), *Sex Tourism and Prostitution: Place, Players, Power, and Politics.* New York: CCC.

Cooper, M. (2000). Tourism in Vietnam: Doi Moi and the Realities of Tourism in the 1990s. In C. M. Hall (Ed.), *Tourism in South and South East-Asia: Issues and Cases.* Oxford: Butterworth-Heinemann.

Craig, J. (1979). *Culture Shock! What Not to Do in Malaysia and Singapore, How and Why Not to Do It.* Singapore: Times Books International.

Crompton, J. L. & MacKay, K. J. (1989). Users' Perceptions of the Relative Importance of Service Quality Dimensions in Selected Public Recreation Programs. *Leisure Sciences,* 11, 367-375.

Crompton, J. L., & Love, L. L. (1995). The Predictive Validity of Alternative Approaches of Evaluating Quality of a Festival. *Journal of Travel Research,* 34(1), 11-24.

Cronbach, L. J. (1951). Coefficient Alpha and the Internal Structure of Tests, *Psychometrica,* 16(3), 297-334.

Czinkota, A. & Ronkainen, I (1993). *International Marketing.* 3rd Ed. Orlando: The Dryden Press.

Damen. L. (1987). *Culture Learning: The Fifth Dimension in the Language Classroom.* Second Language Professional Library. Reading, Mass: Addison-Wesley Publishing Company.

DeVellis, R. F. (1991). *Scale Development: Theory and Applications,* California: Sage Publications.

Dimanche, F. (1994). Cross-Cultural Tourism Marketing Research: An Assessment and Recommendations for Future Studies. *Journal of International Consumer Marketing* 6(3/4): 123-134. In Uysal, M. (ed.) *Global Tourist Behavior.* The Haworth Press, Inc., 123-134.

Dodd, C. H. (1987). *Dynamics of Inter cultural Communication.* Dubuque, IA: Brown, William, C. Publishers

Douglas, M. (1973). *Natural Symbols: Explorations in Cosmology.* New York: Vintage.

Douglas, M. (1978). *Cultural Bias.* Occasional Paper No. 35. London: Royal Anthropological Institute of Great Britain and Ireland.

Edelstein. A. S., Ito. Y. & Kepplinger. H. M. (1989). *Communication and Culture: A Comparative Approach.* New York: Longman.

Ember. C. R. & Ember. M. (1985). *Anthropology.* 4th Ed. New York: Englewood Cliffs. Prentice-Hall.

Embassy of France in Vietnam. (2018 25 January). *Visit to Vietnam by Jean-Baptiste Lemoyne* https://www.diplomatie.gouv.fr/en/country-files/vietnam/.

Engel, J. F. (1995). *Towards the Conceptualisation of Consumer Behaviour*, in J. N. Sheth & T. C. Tan (Eds.). *Conference Proceedings: Historical and International Perspectives of Consumer Research*, Singapore: National University of Singapore and Association for Consumer Research, 1-4.

English Tourist Board (1978). Study of Londoners' Attitudes to Tourists. *Journal of Tourist Research,* 17, 19.

Euromonitor (2002). *European Travel Monitor. Global Market Information Database: Travel and Tourism in France* (Published: September 2002).

Euromonitor International (2003). *Travel and Tourism in USA: Executive Summary* August 2003.

Feather, N. T. (1975). *Values in Education and Society.* New York: Free Press.

Feather, N. T. (1980a). Similarity of Values Systems within the Same Nation: Evidence from Australia and Papua New Guinea. *Australian Journal of Psychology,* 32(1): 17-30.

Feather, N. T. (1980b). Value Systems and Social Interaction: A Field Study in a Newly Independent Nation. *Journal of Applied Social Psychology,* 10(1), 1-19.

Feather, N. T. (1994). Values and Culture. In *Psychology and Culture,* Lonner, W. J. & Malpass, R. S. (Eds.). Boston: Allyn and Bacon.

File, K. M., Judd. B. B. & Prince, R. A. (1992). Interactive Marketing: The Influence of Participation on Positive Word-of-Mouth and Referrals. *The Journal of Service Marketing,* 6 (4), 5-14.

Finn, A. & Kayande, U. (1997). Reliability Assessment and Optimisation of Marketing Measurement. *Journal of Marketing Research,* 34(May), 262-275.

Fontaine, G. (1983). Americans in Australia: Intercultural Training for the Lucky Country. In Landis. D. and Brislin. R. (Eds.) *Handbook of Intel-Cultural Training 3.* New York: Pergamon Press.

Fridgen, J. D. (1991). *Dimensions of Tourism.* USA: The Educational Institute of the American Hotel and Motel Association.

Gagliano, K. B. & Hathcote, J. (1994). Customer Expectations and Perceptions of Service Quality in Apparel Retailing. *Journal of Services Marketing,* 8 (1), 60-69.

Gee, C., Makens, J. & Choy, D. (1989). *The Travel Industry.* New York: Van Nostrand Reinhold. Geertz, C. (1973). Thick Description: Towards an Interpretative Theory of Culture, in *The Interpretation of Cultures: Selected Essays,* New York: Basic Books, 3-32.

Goodenough, W. H. (1981). *Culture, Language and Society.* 2 Ed., Menlo Park: Benjamin Benjamin and Cummings Publication Company.

Goodenough, W. H. (1971). Culture, Language and Society. *Anthropology 7.* Reading, Mass, Addison-Wesley.

Goffman, E. J. (1961). *Encounters, Two Studies in the Sociology of Interaction.* New York: Bobbs-Merrill Co.

Gorsuch, R. L. (1983). *Factor Analysis,* 2nd Ed, Hillsdale, New Jersey: Laurence Eribaum Associates.

Grönroos, C. (1982). A Service Quality Model and Its Marketing Implications. *European Journal of Marketing,* 18(1), 36-4.

Grönroos, C. (1990). *Service Management and Marketing.* Lexington: Lexington Books.

Gudykunst, W. B. (1988). *Theories in Intercultural Communication.* Newbury Park, CA: Sage Publications.

Gudykunst, W. B. & Kim, Y. Y. (1984a). *Communicating with Strangers: An Approach to Intercultural Communication.* Reading, Mass.: Addison-Wesley Publisher.

Gudykunst, W. B. & Kim, Y. Y. (1984b). *Methods for Intercultural Communication Research.* Beverly Hills, CA: Sage Publications.

Hair, J. F. J, Anderson, R. E., Tatham, R. L., & Black, W. C. (1995). *Multivariate Data Analysis with Reading,* 4th Ed, Englewood Cliff, NJ.: Prentice Hall.

Hair, J. F. J, Anderson, R. E., Tatham, R. L., & Black, W. C. (1998) *Multivariate Data Analysis,* 5th Ed. Upper Saddle River, NJ.: Prentice Hall International.

Hall. E. T. (1955). The Anthropology of Manners. *Scientific American,* 192(4): 84-88.

Hall, E. T. (1959/1973). *The Silent Language.* Garden City, New York: Doubleday and Fawcett Company, Anchor Press.

Hall, E. T. (1960). The Silent Language in Overseas Business. *Harvard Business Review,* May-June.

Hall, E. T. (1966). *The Hidden Dimension.* Garden City, New York: Doubleday and Fawcett Company.

Hall, E. T. (1976/1977). *Beyond Culture.* New York: Doublcday, Anchor Press.

Hall, E. T. (1983). *The Dance of Life: The Other Dimensions of Time.* New York: Doubleday.

Hall, E. T. & Hall, M. R. (1987). *Hidden Differences: Doing Business with the Japanese.* Garden City, New York: Anchor Press/Doubleday.

Hampden-Turner, C. & Trompenaars, F. (1993). *Seven Cultures of Capitalism: Value Systems for* Creating Wealth in the United States. Britain, Japan, Germany, France, Sweden and the Netherlands. New York: Doubleday.

Hargie, 0. (1986). *A Handbook of Communication Skills.* London: Routledge.

Harre, R. & Secord, P. (1972). *The Explanation of Social Behavior.* Oxford: Basil Blackwell.

Haring M. D. & Mattsson, J. (1999). A Linguistic Approach to Studying Quality of Face-to-Face Communication. *Service Industries Journal,* 79(2), 28-48.

Harris, P. R. & Moran, R. T. (1996). *Managing Cultural Differences: Leadership Strategies for a New World of Business* (4th Ed.), Houston, TX: Gulf Publishing Company.

Heidi, J. B. & John, G. (1988). The Role of Dependence Balancing in Safe Guarding Transaction-Specific Assets in Conventional Channels. *Journal of Marketing,* 56(April), 32-44.

Herbig, P. A. (1998). *Handbook of Cross-cultural Marketing.* Binghamton, NY: International Business Press.

Hofstede, G. H. (1980). Cultures Consequences: International Differences, in *Work Related Values.* Beverly Hills CA: Sage.

Hofstede, G. H. (1984). The Cultural Relativism of the Quality of Life Concept. *Academy of Management Review,* 9, 389-398.

Hofstede, G. H. (1996). Images of Europe: Past, Present and Future. In Joynt, P. and Warner, M. (Eds.). *Managing Across Cultures: Issues and Perspectives.* North Yorkshire, UK: International Thomson Business Press.

Hofstede, G. H. (1991/1997). *Cultures and Organizations: Software of the Mind.* London/New York: McGraw-Hill, International.

Hofstede, G. H. (2001). *Culture's Consequences: Comparing Values, Behaviours, Institutions and Organizations across Nations.* 2nd Ed. Thousand Oaks: Sage Publications.

Hofstede, G. H. & Bond, M. H. (1984). Hofstede's Culture Dimensions: An Independent Validation of Rokeach's Value Survey. *Journal of Cross-Cultural Psychology*, 15, 417-433.

Hofstede, G. H. & Bond, M. H. (1988). The Confucius Connection: From Cultural Roots to Economic Growth, *Organizational Dynamics,* (16), 5-21.

Hofstede, G. H. & Hofstede, G. J. (2005). *Cultures and Organizations Software of the Mind* (Edition Reviewed and Expanded, 2nd Ed.): Publisher New York: McGraw-Hill.

Horowitz, D. (1985). *Ethnic Groups in Conflict.* Berkeley, CA: University of California Press.

Howat, G., Absher, J, Crilley, G. & Milne, I. (1996). Measuring Customer Service Quality in Sports and Leisure Centres. *Managing Leisure,* 1, 77-89.

Huang, J. H., Huang, C. T. & Wu, S. (1996). National Character and Response to Unsatisfactory Hotel Service. *International Journal of Hospitality Management,* 15(3), 229-243.

Hui, C. H. & Triandis, H. C. (1985). Measurement in Cross-cultural Psychology: A Review and Comparison of Strategies. *Journal of Cross-Cultural Psychology,* 16(2), 131-152.

Hsu, F. L. K. (1971a). Psychosocial Homeostasis and Jen: Conceptual Tools for Advancing Psychological Anthropology. *American Anthropologists,* 73(1), 23-44.

Hsu, F. L. K. (1971b). Filial Piety in Japan and China. *The Journal of Comparative Family Studies,* 2, 67-74.

Hui, C. H. & Triandis, H. C. (1985). Measurement in Cross-cultural Psychology: Review and Comparison of Strategies. *Journal of Cross-Cultural Psychology,* 16(2), 131-152.

Huyton, J. R. (1991). *Cultural Differences in HE: The Example of Hong Kong Students in British Educational Systems/Institutions.* Paper presented at the International Association of Hotel Management Schools Spring 1991 Symposium, East Sussex, Brighton, April 4-5.

Inkeles, A. & Levinson, D. J. (1969). National Character; the Study of Modal Personality and Socio-Cultural Systems. In Lindzey. G. and

Aronson. E. (Eds.). *The Handbook of Social Psychology* 4. Reading. MA: Addison-Wesley: 418-506.

Jacquemin, C. (2001). *France Outbound Market Report*: Pacific Asia Travel Association.

Jansen-Verbeke, M., and Go, F. (1995). Tourism development in Vietnam. *Tourism Management, 16*(4), 315-325.

Jensen, J. (1970). *Perspectives on Oral Communication*. Boston: Holbrook Press.

Jones, P. (1993). *Studying Society: Sociology Theories and Research Practices*, Sociology and Science, Collins Educational, London.

Jones, L. V. & Bock, R. D. (1960). Multiple Discriminant Analysis Applied To A "Way to Live": Ratings From Six Cultural Groups. *Sociometry* 23, 162-176.

Kahle, L. R. (1983). *Social Values and Social Change: Adaptation to Life in America*. New York: Praeger.

Kahle, L. R. (1996). Social Values and Consumer Behavior: Research from the List of Values. In: Seligman, C., Olson, J. M. and Zanna, M. P., (Eds.). *The Psychology of Values:* the Ontario Symposium, Vol. 8, Hillsdale, NJ: Lawrence Eribaum Associates 135-151.

Kamakura, W. A. & Mazzon, J. A. (1991). Value Segmentation: A Model for the Measurement of Values and Value System. *Journal of Consumer Research,* 18(2), 208-218.

Kamakura, W. A. & Novak, T. P. (1992). Value System Segmentation: Exploring The Meaning of LOV, *Journal of Consumer Research,* 19, 119-132.

Kamal, A. A. & Maruyama, G. (1990). Cross-Cultural Contact and Attitudes of Qatari Students in the United States. *International Journal of Inter Cultural Relations,* 14, 123-134.

Keown, C., Jacobs, L. & Worthley, R. (1984). American Tourists' Perceptions of Retail Stores in 12 Selected Countries. *Journal of Travel Research,* 22(3), 26-30.

Kikuchi, A. & Gordon, L. V. (1966). Evaluation and Cross-Cultural Application of a Japanese Form of the Survey of Interpersonal Values. *Journal of Social Psychology,* 69, 85-195.

Kikuchi, A. & Gordon, L. V. (1970). Japanese and American Personal Values: Some Cross-Cultural Findings. *International Journal of Psychology,* 5,183-187.

Kim, K. & Frazier, G. L. (1997). Measurement of Distributor Commitment in Industrial Channels of Distributor. *Journal of Business Research,* 40,139-154.

Kim, Y. Y. & Gudykunst, W. B. (1988). *Theories in Intercultural Communication. International and Intercultural Communication.* Annual 12. Newbury Park, CA: Sage Publications.

Kirkbride, P. S., Tang, S. F. Y. & Westwood, R. I. (1991). Chinese Conflict Preferences and Negotiating Behaviour: Cultural and Psychological Influences. *Organization Studies,* 12, 365-386.

Kline, P. (1994). *An Easy Guide to Factor Analysis,* New York: Routledge.

Kirkbride, P. S., Tang, S. F. Y. & Westwood, R. I. (1991). Chinese Conflict Preferences and Negotiating Behaviour: Cultural and Psychological Influences. *Organization Studies,* 12, 365-386.

Kluckhohn, C. (1944). *Mirror for Man.* New York: McGraw-Hill.

Kluckhohn, C. & Kelly, W. H. (1945). The Concept of Culture. In Linton, R. (ed.) *The Science of Man in the World of Crisis.* New York; Columbia University Press: 78-106.

Kluckhhohn, C. (1951). Values and Value-orientations in the Theory of Action: An Exploration in Definition and Classification, in T. Parsons and E.A. Shils (Eds.) *Toward a General Theory of Action,* New York: Harper & Row.

Kluckhohn, C. (1956). Toward a Comparison of Value-Emphases in Different Cultures. In White, L. D. (Ed.). *The State of the Social Sciences.* Chicago: University of Chicago Press: 116-132.

Kluckhohn, C. (1959). *The Scientific Study of Values.* In University of Toronto Installation Lectures. Toronto: University of Toronto Press.

Kluckhohn, C. & Strodtbeck, F. L. (1961). *Variations in Value Orientations.* New York: Harper and Row.

Ko, A., Chiu, R. & M., W. (1990, December). *Significance of Cultural Change and Its Implications to Management Practices.* Paper presented at the First International Organizational Behaviour Teaching Conference, Singapore.

Kohn, M. L. (1969). Class *and Conformity: A Study in Values.* Dorsey: Homewood.

Kotler, P., Chandler, P., Gibbs, R. & McColl, R. (1989). *Marketing in Australia.* 2nd Ed. New York: Prentice-Hall.

Kroeber. A. & Kluckhohn, C. (1952). *Culture: A Critical Review of Concepts and Definitions.* Papers of the Peabody Museum of American Archaeology and Ethnology, Harvard University Press 47(1): 223. New York: Random House.

Kroeber, A. & Kluckhohn, C. (1985). *Culture: A Critical Review of Concepts and Definitions.* New York: Random House.

Landis, D. & Brislin, R. W. (1983). *Handbook of Intercultural Training 2 and 3: Issues in Training Methodology.* New York: Pergamon Press.

Lau, S. K., & Kuan, H. C. (1988). *The Ethos of the Hong Kong Chinese.* Hong Kong: The Chinese University Press.

Laurent, A. (1980). Once a Frenchman Always A Frenchman. *International Management.* 35(6), 45-46.

Lee, H. S. & Kim, Y. (1999). Service Quality versus Service Value. *The Korean Journal of Marketing,* 1(2), 77-99.

Lehtinen, U. & Lehtinen, J. R. (1982). *Service Quality: A Study of Quality Dimensions.* Working Paper, Service Management Institute, Helsinki Finland.

Leighton, A. H. (1981). Culture and Psychiatry. *Journal of Psychiatry*, 26, 522-529.

Lessem, R. & Neubauer, F. (1994). *European Management Systems: Towards Unity out of Cultural Diversity.* London: Mcdraw-Hili.

Leung, E. (1991). *Cross-Cultural Impacts in Classroom Situations from a Student Perspective.* Paper presented at the International Association of Hotel Management Schools Spring 199] Symposium, East Sussex, Brighton, April 4-5.

Leung, K. (1988). Some Determinants of Conflict Avoidance. *Journal of Cross-Cultural Psychology,* 9(l), 125-136.

Leung, K. & Bond, M. H. (1984). The Impact of Cultural Collectivism on Reward Allocation. *Journal of Personality and Social Psychology,* 47, 793 804.

Leung, K. & Bond, M. H. (1989). On the Empirical Identification of Dimensions for Cross-Cultural Comparisons. *Journal of Cross-Cultural Psychology,* 20, 133-51.

Lewis-Beck, M. (1994). *Data analysis: An introduction,* Sage Publications Thousand Oaks, California.

Lovejoy, A. 0. (1950). Terminal and Adjectival Values. *Journal of Philosophy* 47, 593- 608.

Luk, S., Leon, C., Leon, F., Li, E. & DeLeon, C. (1993). Value Segmentation of Tourists' Expectations of Service Quality. *Journal of Travel and Tourism Marketing,* 2 (4), 23-38.

Luna, D. & Gupta, S. F. (2001). An Integrative Framework for Cross-cultural Consumer Behavior, *International Marketing Review,* 18(1), 45-69.

Lustig, M. W. (1988). Value Differences in Intel-Cultural Communication,' in *Intercultural Communication: A Reader,* Ed. Samovar, L. A and Porter, R. E., Wadsworth Inc., California.

Lustig, M. & Koester, J. (1993/1999). *Intercultural Competence: Interpersonal Communication across Cultures.* New York: Harper Collins.

Lynch, F. (1970). Social Acceptance Reconsidered. In Lynch F. and de Guzman (Eds.) *Four Readings on Philippine Values.* 3rd Ed. Quezon City: Ateneo de Manila University Press.

Lovelock, C. H. (1991). *Services Marketing.* 2nd Ed. New Jersey: Prentice Hall. Madrigal, R. & Kahle, L. (1994). Predicting Vacation Activity Preferences on the Basis of Value-System Segmentation. *Journal of Travel Research,* 32 (3), 22-28.

Maholtra, N. K. (1996). *Marketing Research: An Applied Orientation,* Englewood, Cliff, NJ: Prentice Hall

Martin, W. B. (1987). A New Approach to the Understanding and Teaching of Service Behavior. *Hospitality Education and Research Journal,* 11(2), 255-262.

Mayo, E. J. & Jarvis, L. P. (1981). *The Psychology of Leisure Travel.* CBI, Boston.

Maslow, A. H. (1943). A Theory of Human Motivation. *Psychological Review, 50,* 370-396.

Maslow, A. H. (1959). *New Knowledge in Human Values.* New York: Harper.

Maznevski, M. (1994). *Synergy and Performance in Multi-cultural Teams.* PhD Thesis University of Western, Ontario.

McCort, D. J., & Malhotra, N. K. (1993). Culture and Consumer Behaviour: Toward an Understanding of Cross-Cultural Consumer Behaviour in International Marketing. *Journal of International Consumer Marketing,* 6(2), 91-127.

McIntosh R. W. & Goelner C. R. (1986*). Tourism: Principles, Practices, Philosophies,* John Wiley and Sons, Inc; New York.

McCracken, G. (1986). Culture and Consumption: A Theoretical Account of the Structure and Movement of the Cultural Meaning of Consumer Goods. *Journal of Consumer Research,* 13, 71-84.

Mead, R. (1998). *International Management.* 2nd Ed. Oxford: Blackwell.

Meissner, S. J. (1971). Notes toward a Theory of Values: Values as Cultural. *Journal of Religion and Health,* 10, 77-97.

Milbraith, L. (1980). *Values, Lifestyles and Basic Beliefs as Influences on Perceived Quality of Life.* Paris: UNESCO.

Mill, R. C. & Morrison, A. M. (1985). *The Tourism System: An Introductory Text. Englewood Cliffs*, New Jersey: Prentice Hall.

Millington, K. (2001). Vietnam: Country Report. *EIU Travel and Tourism Analysis*, 2, 87- 97.

Mirels. H. L. & Garrett, J. B. (1971). The Protestant Ethic as a Personality Variable. *Journal of Consulting and Clinical Psychology,* 36, 40-44.

Morris, C. W. (1956). *Varieties of Human Value.* Chicago: University of Chicago Press.

Mourn, T. (1980). *The Role of Values and Life-Goals in Quality of Life.* Paris: UNESCO.

Moutinho, L. (1987). Consumer Behaviour in Tourism. European. *Journal of Marketing,* 21(10), 5-44.

Mowen, J. C. (1993/1995). *Consumer Behaviour* (4th Ed.). Englewood Cliffs, N.J.: Prentice-Hall.

Nhan Dan (2018, March 25). *Deepening Vietnam-France Strategic Partnership.* https://en.nhandan.org.vn/politics/editorial/item/6792402-deepening-vietnam-france-strategic-partnership.html.

Newman, W. L. (1997). *Social Research Methods: Qualitative and Quantitative Approaches*, Allyn and Bacon, Boston.

Nieto, S. (1996). *Affirming Diversity: The Socio Political Context of Multicultural Education.* (2nd Ed.) New York: Longman.

Nomura, N. & Bamlund. D. (1983). Patterns of Interpersonal Criticism in Japan and the United States. *International Journal of Intercultural Relations,* 7(1), 1-18.

Norusis, M. J. (1993). *SPSS for Windows, Professional Statistics,* Release 6.0, Chicago: SPSS Inc.

Nunnally, J. C. (1967/1978). *Psychometric Method,* New York: McGraw-Hill Book Company.

Ohio State Parks (1996). *Improving Customer Satisfaction with Parks and Recreation Services under Smaller Budgets.* Unpublished Technical Report. Columbus, OH: Ohio State Parks.

Oliver, R. L. (1980). A Cognitive Model of the Antecedents and Consequences of Satisfaction Decisions. *Journal of Marketing Research,* 17(4), 460-470.

Parasuraman, A., Zeithaml, V. A. & Berry, L. L. (1985). A Conceptual Model of Service Quality and its Implications for Future Research. *Journal of Marketing,* 49(Fall), 41-50.

Parasuraman, A., Zeithaml, V. A. & Berry, L. L. (1988). SERVQUAL: A Multiple-Item Scale for measuring Consumer Perceptions of Service Quality. *Journal of Retailing,* 64 (Spring), 12-37.

Parasuraman, A., Zeithaml, V. A. & Berry, L. L. (1991). Refinement and Reassessment of the SERVQUAL Scale. *Journal of Retailing,* 67(4), 420-450.

Parasuraman, A., Zeithaml, V. A. & Berry, L. L. (1994). A Parasuraman, V. A. Zeithaml and L. L. Berry, Reassessment of Expectations as a

Comparison Standard in Measuring Service Quality: Implications for Further Research, *Journal of Marketing,* 58, 120–135.

Parsons, T. (1951/1953). *The Social System,* Glencoe, IL: Free Press.

Parsons, T. (1973). Culture and Social System Revisited in *The Idea of Culture in the Social Sciences,* Eds. L. Schneider and C. Bonjean, London: Cambridge University Press, 33-46.

Parsons, T. & Shils, E. A. (1951). *Toward a General Theory of Action.* Cambridge, MA: Harvard University Press.

Patterson, P. G. (2000). A Contingency Approach to Modelling Satisfaction with Management Consulting Services. *Journal of Service Research,* 3(2), 138 -153.

Patterson, P. G., & Johnson, L. W. (1993). Disconfirmation of Expectation and the Gap Model of Service Quality: An Integrated Paradigm. *Journal of Consumer Satisfaction, Dissatisfaction and Complaining Behaviour,* 6, 90-99.

Patterson, P. G., Johnson, L. W. & Spreng, R. A. (1997). Modeling the Determinants of Customer Satisfaction for Business to-Business Professional Services. *Journal of the Academy of Marketing Science,* 25 (1), 4-17.

Pearce, P. L. (1982a). Perceived Changes in Holiday Destination. *Annals of Tourism Research*, 9 (1): 145-164.

Pearce, P. L. (1982b). *The Social Psychology of Tourist Behaviour.* International Series in Experimental Social Psychology. Vol. 3. Oxford; New York: Pergamon Press.

Pearce, P. L. & Moscardo, G. M. (1984). Making Sense of Tourists' Complaints, *Tourism Management*, 5(1), 20-23.

Peter, P. J. (1979). Reliability: A Review of Psychometric Basics and Recent Marketing Practices. *Journal of Marketing Research,* 16(1), 6-17.

Peter, P. J. (1981). Construct Validity: A Review of Basic Issues and Marketing Practices. *Journal of Marketing Research,* 18(2), 133-145.

Peterson, R. A. (1979). Revitalizing the Culture Concept. *Annual Review of Sociology,* 5, 137-165.

Pizam, A. (1978). Tourism's Impacts: The Social Costs to the Destination Community as Perceived by its Residents. *Journal of Travel Research*, 16(4), 8-12.

Pizam, A. (1999). The American Group Tourist as Viewed by British, Israeli, Korean, and Dutch Tour guides. *Journal of Travel Research*, 38(2), 119 -126.

Pizam, A. (1999). Cross-cultural Tourist Behaviour. In: Pizam, A. and Mansfeld, Y., Editors, 1999. *Consumer Behaviour in Travel and Tourism* (393-411). New York: Haworth Press.

Pizam, A., Neumann, Y. & Reichel, A. (1978). A Pizam, Y. Neumann and A. Reichel, Dimensions of Tourism Satisfaction with a Destination Area, *Annals of Tourism Research* 5, 314–322.

Pizam, A. & Telisman-Kosuta, N. (1989). Tourism as a Factor of Change: Results and Analysis. In J. Bystrzanowski (Ed.), *Tourism as a Factor of Change: A Socio-Cultural Study 1,* (149-156). Vienna: European Coordination.

Pizam, A., & Sussmann, S. (1995). Does Nationality Affect Tourist Behavior? *Annals of Tourism Research,* 22(4), 901-917.

Pizam, A., Neumann, Y. & Reichel, A. (1996). The Effect of Nationality on Tourist Behaviour: Israeli Tour-Guides Perception. *Journal of Hospitality and Leisure Marketing,* 4(1), 23-49.

Pi-Sunyer, O. (1978). Through Native Eyes: Tourists and Tourism in Catalan Maritime Community. In V. Smith, L. (Ed.). *Host and Guests: The Anthropology of Tourism*. Philadelphia: University of Pennsylvania Press.

Pi-Sunyer, O. (1982). The Cultural Costs of Tourism. *Cultural Survival Quarterly* 6(3), 7.

Pookong, K., and King, B. (Eds.). (1999). *Asia-Pacific Tourism Regional Co-operation Planning and Development* (1st ed.). Melbourne: Hospitality Press.

Porter. R. E. & Samovar, L. A. (1988) Approaching Intercultural Communication. In Samovar. E. A. and Porter, R. E. (Eds.). *Intercultural Communication: A Reader.* 5th Ed. Belmont. CA: Wadsworth Publishing Company.

Potter, C. C. (1989).What is Culture: And Can It Be Useful for Organizational Change Agents? *Leadership and Organization Development Journal,* 10 (3), 229-237.

Redding, G. S. (1980). Management Education for the Orientals. In *Breaking Down Barriers Practice and Priorities or International Management Education,* B. Garratt and J. Standford, Eds., Surrey, Great Britain: Gower Publishing, 193-214.

Redding, G. S. (1982). Cultural Effects on the Marketing Process in Southeast Asia. *Journal of Marketing Research Society,* 24, 98-114.

Redding, G. S. (1990). *The Spirit of Chinese Capitalism,* NY: Walter de Gruytur.

Reisinger, Y. & Turner, L. (1998). Cross-cultural Differences in Tourism: A Strategy for Tourism Marketers. *Journal of Travel and Tourism Marketing,* 7(4), 79-106.

Reisinger, Y. & Turner, L. (2002a). Cultural Differences between Asian Tourist Markets and Australian Hosts Part 1. *Journal of Travel Research,* 40(3), 295-315.

Reisinger, Y. & Turner, L. (2002b). Cultural Differences between Asian Tourist Markets and Australian Hosts, Part 2. *Journal of Travel Research,* 40(4), 74-384.

Reisinger, Y. & Turner, L. W. (2003). *Cross-cultural Behaviour in Tourism: Concepts and Analysis.* Oxford: Butterworth-Heinemann.

Richardson, S. L. & Crompton, J. L. (1988). Cultural Variations in Perceptions of Vacation Attributes, *Tourism Management*, 9 (2), 128-136.

Richter, L. K. (1983). Political Implications of Chinese Tourism Policy. *Annals of Tourism Research,* 10, 347-362.

Robertson, T. S. (1970). *Consumer Behaviour.* Glen view, Ill: Scott: Foresman and Company.

Robinson, G. L. N., & Nemetz, L. (1988). *Cross-Cultural Understanding.* UK: Prentice Hall International.

Rokeach, M. (1968a). A Theory of Organization and Change within Value-Attitude Systems. *Journal of Social Issues,* 24: 13-33.

Rokeach, M. (1968b). *Beliefs, Attitudes and Values*. San Francisco: Jossey-Bass.

Rokeach, M. (1971). Long-Range Experimental Modification of Values, Attitudes and Behavior. *American Psychologist,* 26, 453-459.

Rokeach, M. (1973). *The Nature of Human Values*. New York: Free Press.

Rokeach, M. (1979a) *Understanding Human Values: Individual and Societal.* New York: Free Press.

Rokeach, M. (1979b). From Individual to Institutional Values: With Special Reference to the Values of Science. In M. Rokeach (Ed.), *Societal, Institutional and Organizational Organizational Values* (Vol. 47-70). New York: The Free Press.

Ryan, A. S. (1985). Cultural Factors in Casework with Chinese - Americans. *Social Casework,* 66(6), 333-340.

Ryan, C. (1995). *Researching Tourist Satisfaction: Issues, Concepts, Problems.* Routledge, London and New York.

Salant, P. & Dillman, D. A. (1994). *How to Conduct Your Own Survey*. New York: John Wiley & Sons, Inc.

Saleh, F., & Ryan, C. (1992). Client Perceptions of Hotels: A Multi-Attribute Approach. *Tourism Management,* 13(2), 163-168.

Samovar, L. A., Porter, R. E. & Jain, N. C. (1981). *Understanding Intercultural Communication.* Belmont, CA: Wadsworth Publishing Company.

Samovar, L. A. & Porter, R. E. (1988). *Intercultural Communication: A Reader.* 5th Ed. Belmont, CA: Wadsworth Publishing Company.

Samovar, L. A. & Porter, R. E. (1991). *Communication between Cultures.* Belmont, CA: Wadsworth Publishing Company.

Samovar, L., Porter, R. & Stefani, L. (1998). *Communication between Cultures.* Belmont, CA: Wadsworth Publishing Company.

Sapir, E. (1949/1964). *Culture, Language, and Personality: Selected Essays.* Mandelbaum, D. G. (ed.). Berkeley and Los Angeles, CA: University of California Press.

Schiffman, L. G. & Kanuk, L. L. (1991). *Consumer Behaviour* (4th Ed.), Englewood Cliffs, N.J.: Prentice-Hall Inc.

Schneider, M. & Jordan, W. (1981). Perceptions of the Communicative Performance of Americans and Chinese in Intercultural Dyads. *International Journal of Intercultural Relations,* 5(1), 175-191.

Schneider, S. & Barsoux, J. L. (1997). *Managing Across Cultures.* New York: Prentice-Hall.

Schwartz, S. (1992). Individualism-Collectivism: Critique and Proposed Refinements. *Journal of Cross-Cultural Psychology,* 21, 139-157.

Schwartz, S. (1994). Are there Universal Aspects of the Structure and Content of Values? *Journal of Social Issues,* 50(4), 19-45.

Scollon, R. & Scollon, S. (1995). *Intel-cultural Communication: A Discourse Approach.* Cambridge, MA: Blackwell.

Scott, W. A. (1965). *Values and Organizations: A Study of Fraternities and Sororities.* Chicago: Rand McNally.

Sechrest, L., Fay, T., Zaidi, H. S. (1972) Problems of Translation in Cross-Cultural Research. *Journal of Cross-Cultural Psychology,* 3, 41-56.

Segall, M. H. (1979). *Cross-cultural Psychology: Human Behaviour in Global Perspective.* Monterey, Calif., Brooks /Cole Pub. Co.

Segall, M. H., Dasen, P. R., Berry, J. W. & Poortinga, Y. H. (1990). *Human Behaviour in Global Perspective: An Introduction to Cross-Cultural Psychology.* London: Pergamon Press.

Schmidt, C. J. (1979). The Guided Tour, *Urban Life,* 7, 441-467.

Sheldon, P. J. & Fox, M. (1988). The Role of Foodservice in Vacation Choice and Experience: A Cross-Cultural Analysis. *Journal of Travel Research,* 27(3), 9 -15.

Sikula (1970) in Rokeach, M. (1973). *The Nature of Human Values.* New York: Free Press.

Smart, N. (1968). Attitudes towards Death in Eastern Religions. In Toynbee, A. (ed.) *Man's Concern with Death.* London: Hodder and Stoughton.

Smith, B. (1969). *Psychology and Human Values.* Chicago: Aldine.

Smith. V. L. (1989). *Hosts and Guests: The Anthropology of Tourism.* 2nd Ed. Philadelphia: University of Pennsylvania Press.

Solomon, M. R., Surprenant, C., Czepiel, J. A., & Gutman, E. G. (1985). A Role Theory Perspective on Dyadic Interactions: The Service Encounter. *Journal of Marketing,* 49(1), 99-111.

Sparks, B. (1994). Communicative Aspects of the Service Encounter. *Hospitality Research Journal*, 17, 39-50.

Sparks, B. & Callan, J. (1992). Communication and the Service Encounter: The Value of Convergence. *International Journal of Hospitality Management*, 11, 213-224.

Sudman, S. (1976). *Applied Sampling*. New York: Academic Press.

Sutton, W. A. (1967). Travel and Understanding: Notes on the Social Structure of Touring. *International Journal of Comparative Sociology*, 8(2), 218-223.

Sutton, R. I. & Rafaeli, A. (1988). Untangling the Relationship between Displayed Emotions and Organizational Sales: The Case of Convenience Stores. *Academy of Management Journal*, 31, 461-487.

Stewart, R. A. (1971). Cross-Cultural Personality Research and Basic Cultural Dimensions through Factor Analysis. *Personality*, 2, 45-72.

Stewart, E. C. (1972). *American Cultural Patterns: A Cross-Cultural Perspective*. Chicago: Intercultural Press.

Stoetzel, J. (1955). *Without the Chrysanthemum and the Sword*. New York: Columbia University Press.

Svensson, G. (2001). The Quality of Bi-Directional Service Quality in Dyadic Service Encounters. *Journal of Services Marketing*, 15(5), 357-378.

Tabachnik, B. G. & Fidell, L. S. (1989). *Using Multivariate Statistics*. 2nd Ed. California State University, Northridge: Harper Collins Publishers.

Taylor, H. M. (1974). Japanese Kinesics. *Journal of the Association of Teachers of Japanese*, 9, 65-75.

Terpstra, V. & David, K. (1985). *The Cultural Environment of International Business*. Dallas: South-Western Publishing Co.

Theodorson, G. A. (1969). *A Modern Dictionary of Sociology*. New York, Crowell.

The Jakarta Post (2018, 4 November). France, Vietnam sign $10 billion deals. https://www.thejakartapost.com/seasia/2018/11/04/france-vietnam-sign-10-billion-deals.html.

Thiederman, S. (1989). Overcoming Cultural and Language Barriers. *Public Management* 71, 19-21.

Theuns, H. (1997). Vietnam: Tourism in an Economy in Transition. In F. Go and C. Jenkins (Eds.), *Tourism and Economic Development in Asia and Australasia*. London: Pinter.

Tjosvold, D., Hui, C. & Law, K. S. (2000). Constructive conflict in China: Cooperative Conflict as a Bbridge between East and West, *Journal of Word Business*, 36 (2), 166-183.

Triandis, H. C. (1972). *The Analysis of Subjective Culture*. New York: Wiley-Interscience.

Triandis, H. C. (1977a) Subjective Culture and Interpersonal Relations across Cultures. In Loeb-Adler, L. (Ed.). Issues in Cross-Cultural Research. *Annals of the New York Academy of Sciences*, 285, 418-434.

Triandis, H. C. (1977b). *Interpersonal Behavior*. Monterey, CA: Brooks/Cole Publishing Company.

Triandis, H. C. (1979). Values, Attitudes, and Interpersonal Behavior, in *Nebraska Symposium on Motivation* 1979, Ed. H. E. Howe and M. M. Page, Lincoln: University of Nebraska Press, 195-259.

Triandis, H. C. (1990). Cross-Cultural Studies of Individualism-Collectivism, in *Nebraska Symposium on Motivation,* 35, ed. J- Berman, Lincoln, NE: University of Nebraska Press, 41-134.

Triandis, H. C., Kilty, K. M., Shanmugam, A. V., Tanaka, Y. & Vassiliou, V. (1972a). Cognitive Structures and the Analysis of Values. In Triandis, H. C. (Ed.). *The Analysis of Subjective Culture*. New York: John Wiley and Sons.

Triandis, H. C., Vassiliou, V., Vassiliou, G., Tanaka, Y. & Shanniugam, A. V. (1972b). *The Analysis of Subjective Culture*. New York: John Wiley and Sons.

Trompenaars, F. (1984). *The Organisation of Meaning and the Meaning of Organisation - A Comparative Study on the Conceptions and Organisational Structure in Different Cultures*. PhD Thesis, University of Pennsylvania.

Trompenaars, F. (1993/1997). *Riding the Waves of Culture: Understanding Cultural Diversity in Business*. London: Brealey.

Truong, T. H. (2007). *Holiday Satisfaction in Vietnam _Cross-Cultural Perspectives* PhD Dissertation, Victoria University, Australia.

Turner, L. W. (1991). *Business Statistics,* Part A, Department of Applied Economics, Melbourne: Victoria University.

Urriola, 0. (1989). Culture in the Context of Development. *World Marxist Review,* 32, 66-69.

Urry, J. (1991). The Sociology of Tourism. In C. P. Cooper (Ed.), *Progress in Tourism, Recreation and Hospitality Management,* 3(48-57). England: The University of Surrey.

US Central Intelligence Agency. (2019, January 10). *The World factbook— France.* Retrieved March 12, 2019, from http://www.cia.gov/cia/publications/factbook/geos/fr.html.

Van Raaij, W. F. (1978). Cross-cultural Research Methodology, A Case of Construct Validity. *Advances in Consumer Research.* Association for Consumer Research (5) 693-701.

Van Raaij, W. F. & Francken, D. A. (1984). Vacation Decisions, Activities and Satisfaction. *Annals of Tourism Research,* 11(1), 101-112.

Veal, A. J. (1998). *Research Methods for Leisure and Tourism: A Practical Guide,* Financial Times, Pitman Publishing, London.

Venkatesh, A. (1995). Ethnoconsumerism: A New Paradigm to Study Cultural and Cross-cultural Consumer Behaviour, in J. A. Costa & G. J. Bamossy (Eds.), *Marketing in a Multicultural World: Ethnicity, Nationalism and Cultural Identity,* SAGE Publications, Thousand Oaks, California.

VNAT. (2019). Vietnam National Administration for Tourism*: Tourist Statistic Visitors. Available from* www.vietnamtourism.com.

Wagatsuma, H. & Rosett, A. (1986). The Implications of Apology: Law and Culture in Japan and the United States. *Law and Society Review* 20(4): 461-498.

Wagner, U. (1977). Out of Time and Place - Mass Tourism and Charter Trips. *Ethnos, 42,* 38-52.

Wallendorf, M. & Reilly, M. (1990). Distinguishing Culture of Origin from Culture of Residence. *Advances in Consumer Research, 10, 699-701.*

Wallerstein, I. (1990). Culture as the Ideological Battle Ground of the Modem World System. In Featherstone, M. (Ed.) *Global Culture -*

Nationalism, Globalization and Modernity. London, Sage Publications, pp. 31-57.

White, L. A. (1959). Concept of Culture. *American Anthropologist* 61: 227-251.

White, R. K. (1951). *Value Analysis: Nature and Use of the Method.* Arbor, A. Michigan: Society for the Psychological Study of Social Issues.

Wei, L., Crompton, J. L. & Reid, L. M. (1989). Cultural Conflicts: Experiences of US Visitors to China. *Tourism Management,* 10(4), 322-332.

Williams, R. M. Jr. (1968). *The Concept of Value.* In Sills, D. (Ed.) Encyclopedia of the Social Sciences. U.S.: MacMillan and Free Press.

Williams, R. M. Jr. (1978). *Values.* International Encyclopedia of the Social Sciences. New York: Crowell Gollier and Macmillan.

Williams, R. M. Jr. (1979). Change and Stability in Values and Value Systems: A Sociological Perspective. In Rokeach, M. (ed.) *Understanding Human Values: Individual and Societal.* New York: Free Press.

Wolfgang, A. (1979). *Nonverbal Behavior: Applications and Cultural Implications*. New York: Academic Press.

Woodside, A. G. & Lysonski, S. (1989). *A General Model of Traveler Destination.*

World Tourism Organization. (2005, October). *Tourism highlights* (2005 Edition). Retrieved February 12, 2006, from http://www.world-tourism.org/facts/eng/pdf/high lights/2005_eng_high.pdf.

Yates, J. F., Lee, J.-W., & Shinotsuka, H. (1996). Beliefs about overconfidence, including its cross-national variation. *Organizational Behavior and Human Decision Processes, 65*(2), 138–147. https://doi.org/10.1006/obhd.1996.0012.

Yau, 0. H. M. (1994). *Consumer Behaviour in China: Customer Satisfaction and Cultural Values.* London: Routledge.

Zavalloni, M. (1980). Values. In Triandis, H. C. and Brislin, R. W. (eds.) *Handbook of Cross-Cultural Psychology: Social Psychology 5.* Boston, MA: Allyn and Bacon.

ABOUT THE AUTHOR

Thuy-Huong Truong
Lecturer & Research Fellow,
Victoria University Business School Victoria University

As a Lecturer, Dr. Thuy-Huong Truong has convening and teaching a range of subjects from marketing, tourism, hospitality, management, consumer behaviour and cross-cultural study spanning over 25 years at various institutions in Australia, Vietnam and France. Furthermore, she has also coordinating and teaching numerous subjects at many offshore campuses including Malaysia, Hong Kong, Kuwait, China, France and Vietnam.

As a Researcher, Dr. Thuy-Huong has publishing many books, book chapters and academic journal articles over the past 18 years. Her research journey has been rewarding too with a highlight of various conferences in Australia, Hong Kong, China, Indonesia, New Zealand, Taiwan and Vietnam.

Dr. Thuy-Huong's research interests and areas of specialisation covering from cross-cultural study, international marketing and tourism, marketing services and experiences, consumer behaviour, destination resources and product development, tourism in the Asia Pacific regions and developing countries.

Additionally, with her multiple language skills and extensive industry experiences in marketing and tourism, Dr. Thuy-Huong Truong was an assistant manager and a business consultant for numerous international joint-venture companies and organisations across Vietnam. These comprise of various fields from tourism, marketing, hospitality, medicine, engineering and construction, water and wastewater, forestry and agriculture.

INDEX

#

18 terminal and 18 instrumental values, 69

A

abundant historical and cultural heritage, 1
accommodation facilities, 5
achieve uniformity, 50
achievement cultures, 101
achievement oriented, 37, 118
activity orientation dimension, 112
actual performance, 48
adaptive characteristics, 24
aesthetic values, 39
affective, 19, 28, 103, 105, 106
affective and neutral cultures, 105, 106
affective components, 19
affective cultures, 106
affordable destination, 4
Africa, 35
after consumption, 50
American culture, 99, 103, 104, 107
analytical procedures, 61

Anglo-Saxon, 32, 33
Anglo-Saxon cultures, 32
anthropological literature, 13
anthropologists, 29
anthropology, 14, 132, 133, 134, 136, 144, 147
apologetic nature, 34
appropriate equivalence of measures, 62
appropriate promotional strategies, 126
appropriate tourism organisation, 5
artifacts, 13
ascription cultures, 101
Asian and Western nationalities, 126
Asian and Western source markets, 125
Asian country, 2
Asian cultures, vii, 8, 70, 104, 107
Asian destinations, vii, 9
Asian societies, 114
assessment index, 124
assurance, 46, 89, 91, 93, 94
asymmetrical reciprocation, 104
attentiveness, 51, 90, 92, 93, 94, 96, 111, 116

attitudes, 3, 5, 14, 15, 16, 20, 21, 24, 25, 26, 28, 30, 31, 38, 40, 44, 51, 55, 57, 61, 62, 63, 112, 130, 132, 137, 146, 147, 149
attitudes as encounter, 25
attractions, 2, 4, 67, 68, 70, 126
attributes and performance of service providers, 55
authority, 99, 100, 101, 119
authority systems, 101
avoidance of conflict, 34
avoiding complaint, 102
avoiding embarrassment, 31, 70

B

Bao, 104
behaviour and attitude, 55
behaviour patterns, 20, 38
behavioural characteristics, 40, 57
behavioural components, 28
behavioural consequence, 15
behavioural differences, 10, 24
behavioural dimension, 75, 81, 89, 121
behavioural norms, 30
beliefs, 14, 15, 16, 20, 21, 23, 24, 29, 31, 38, 44, 55, 69, 108, 115, 141, 146, 151
beliefs and values, 14, 15
belongingness, 35
birthdays of others, 103
bureaucratic pyramid, 100
business culture, 17

C

capability, 91, 93, 94, 95, 97, 111, 118, 119
capable and trustworthy, 113
capitalize, 2
Catholicism, 33
characteristics of culture, 13
children, 103, 107
China, 2, 34, 35, 107, 109, 136, 149, 151

Chinese culture connection, 27, 30, 129, 131
Christian religious philosophy, 33
clarity of expression, 86, 87, 105, 125
class, 1, 15, 18, 31, 101, 103, 139
clustering of societies, 26
coefficient alpha, 74, 131
cognitive, 19, 20, 23, 28, 53, 118, 142, 149
cognitive function, 23
cognitive system, 20
collaterality, 102
collectivism, 30, 140
collectivist and hierarchal culture, 119
common beliefs, 30
common coexistence, 33
communal consensus, 102, 108
communication, 14, 16, 31, 33, 43, 86, 105, 106, 109, 110
communication behaviour, 43, 86, 88, 89, 105, 125
communication patterns, 109
communication practices, 14
communication services, 110
communication skills, 110
communication styles, 31, 105
communication type, 33
communicative, 95, 96, 111, 113, 114, 119, 147, 148
communitarian cultures, 102
community, 14, 17, 35, 100, 107
comparable or equivalent across cultures, 62
comparative approach, 61, 132
competitive advantage, 124
competitive edge, 35
compliment others, 106
concept of self, 35
conflict, 16, 19, 20, 21, 29, 34, 35, 37, 99, 102, 149
Confucianism, 33, 35
Confucius, 136
consensus, 102, 108
consideration for others, 86

consistent service, 5
conspicuous consumption, 108
construct across cultures, 73
construct validity, 74, 143, 150
consumption, 10, 108
consumption of commodities while on vacation, 54
context cultures, 37
contextual characteristics, 43
contextual influence, 43
convenience of sample selection, 66
convenience random sample, 67
conviviality of service providers, 42
corporate culture, 17, 129
courtesy & friendliness, 91, 92, 95, 96, 111
criterion validity, 74
criticizing, 34
Cronbach alpha, 74
cross-cultural component, viii
cross-cultural contact, 16, 137
cross-cultural differences, 32, 37, 69
cross-cultural exchanges, 5
cross-cultural experiences, 10, 114
cross-cultural interactions, 43, 44
cross-cultural misunderstanding, 54
cross-cultural perceptions and satisfaction, 16, 110
cross-cultural psychology, 62, 129, 131, 136, 139, 140, 147, 151
cross-cultural research, 39, 55, 61, 62, 65, 69, 130, 147, 149
cross-cultural service encounter, 5
cross-cultural social or professional interaction, 44
cross-cultural study, 65, 153
cross-cultural style of operation, 2
cross-cultural tourism context, 124
cross-cultural understanding, 114, 124, 145
cross-cultural variations in the rules, 70
cross-national exploration, 53, 58
cultural backgrounds, ix, 39, 44, 57, 62
cultural beliefs, 14, 15

cultural characteristics of source markets, viii
cultural context, 24, 25, 41, 52, 69, 118
cultural differences, 10, 11, 16, 18, 20, 21, 26, 29, 30, 32, 33, 34, 35, 37, 38, 39, 40, 41, 42, 43, 44, 54, 66, 69, 79, 95, 118, 120, 123, 125, 131, 135, 136, 145
cultural differentiation, 30
cultural dimension, 13, 29, 33, 75, 79, 81, 96, 121, 148
cultural dissimilarities, 54, 56
cultural factors, 57, 123, 146
cultural familiarity, 56
cultural heritage, 2, 13
cultural institutions, 22
cultural marker, 14
cultural norms, 20, 23
cultural orientations, 30, 32, 37, 41
cultural orientations and expectations, 37, 41
cultural patterns, 30, 32, 109, 148
cultural patterns of interaction, 32, 109
cultural perceptions, viii, 10, 42, 53
cultural perspective, 2, 148, 149
cultural similarities, 54, 79
cultural similarity and familiarity, 56
cultural themes, 11
cultural training, 124, 127, 129, 133
cultural understanding, viii, 56
cultural value orientations, 22, 41
cultural values, viii, ix, 3, 6, 10, 13, 17, 18, 21, 23, 26, 27, 30, 32, 38, 39, 42, 43, 44, 53, 58, 62, 63, 69, 74, 75, 78, 79, 80, 81, 82, 83, 84, 95, 99, 109, 112, 120, 123, 124, 125, 126, 151
cultural variability, 30
cultural variations, 22, 44, 64, 68, 145
culture emic and etic, 63
culture or subculture, 26
culture-free, 63, 131
culture-oriented marketing strategies, 125, 126

culture-rich, 63
cultures, vii, viii, 8, 9, 10, 13, 14, 15, 16, 17, 18, 20, 21, 22, 23, 25, 26, 27, 28, 29, 30, 32, 34, 35, 36, 41, 44, 56, 61, 62, 65, 66, 69, 73, 99, 104, 105, 106, 107, 109, 111, 112, 113, 120, 126, 128, 129, 133, 135, 136, 138, 140, 146, 147, 149
cumulative percentages of the variance, 77
customary laws, 100, 115
customary laws and standards, 100, 115
customer normative expectations, 47
customer participation, 42
customer satisfaction, viii, 45, 49, 142, 143, 151
customers, 2, 42, 43, 46, 47, 51, 94, 114, 119
customs, 2, 14, 15, 30, 67, 69, 71, 90, 91, 93, 94, 96, 116, 120
customs and traditions, 14

D

decision-making behaviour, 54
deference to tradition, 99
degree of expressiveness, 31, 36
designing appropriate strategies, ix
desirable and undesirable states of affairs, 22
desirable behaviour, 21, 25
desired end-goals, 28
destination image, 41, 58
destination selection, 54, 55
developed labour market, 3
developing countries, 124
development of strategies, 10
differences, viii, 10, 15, 16, 17, 18, 20, 21, 23, 26, 27, 29, 30, 31, 32, 34, 35, 36, 37, 38, 39, 40, 41, 42, 43, 44, 48, 57, 61, 62, 63, 69, 74, 77, 79, 81, 89, 99, 110, 112, 121, 123, 124, 125, 126, 131, 135, 140, 145

differences between the cultures of Asian and Western, 10
different behavioral characteristics, 40
different cultural backgrounds, 22, 32, 44, 56, 107, 110, 112, 124
different cultural settings, 61
dimensions of importance, 112
directness, 35, 37
disconfirmation judgment, 48
displays of affection, 36
dissatisfaction, viii, 20, 32, 48, 124, 143
distinct managerial approach, 42
divergence, 109
divergence of interactions, 109
diverse cultural backgrounds, 125
diversity, 10
Doi Moi policy, 2
doing/being/becoming cultures, 112
dominant cultures, 17, 18
drawing, 13
during the delivery and consumption of services, 10
dyadic interaction, 41, 42, 147
dyadic interactions, 42, 147
dyadic service encounter, 42, 148

E

East, 4, 30, 35, 131, 136, 139, 149
East Asia, 1, 4
East Asia backpacker route, 4
Eastern, 9, 13, 22, 32, 33, 34, 35, 36, 37, 41, 99, 101, 105, 106, 107, 147
Eastern and Western cultures, 13
Eastern and Western societies, 32, 36, 107
Eastern bloc countries, 9
Eastern societies, 34, 35, 36, 37, 106
Eastern values, 33, 37
economic cooperation between Vietnam and France, 7
economic progress, 35

economic reforms, 3, 104, 107
economic sector, 3
education, 3, 33, 101
effective communication, 43
effective management practices, 3
effective tourism's policy formulation, 5
egalitarian society, 113
egalitarianism, 99, 100, 115
ego-defensive, 19
egoism, 37
Eigenvalue, 76, 83, 89, 91, 93, 94
embarrassment, 34, 36, 88
embodiment in artifacts, 17
emerging industry, 2
emic, 62
emic approach, 62
emic research, 63
emic versus etic approaches, 63
emotion, 88, 106
emotional calm, 33
emotional expression, 106
emotionally restrained and sensitive, 37
empathy, 46, 57
empirical studies, 13, 32, 37
employee attitudes, 5
employees, 3
encounter, 43, 58, 110
enduring belief, 19
entrepreneurship, 3
environment, 43, 102
environments, 1
equal opportunity, 29
equality, 19, 29, 37, 102
equality of people, 37
ethics, 104
ethnic composition, vii, 8
ethnic groups, 2
ethnicity, 18, 150
etic approach, 62
etic methodology, 62
etiquette, 70, 85, 87, 99, 101, 116
etiquette of social behaviour, 100, 116

Europe, 135
European, 5, 6, 7, 10, 32, 35, 90, 125, 132, 134, 139, 141, 144
European culture, 10, 32, 90
European values, 32
everyday life, 100
evidence, viii, 16, 32, 123
examined tourist–host service encounters interaction, viii
exotic, 2, 109
exotic plants, 2
expectancy-disconfirmation paradigm, 48
expectation-performance confirmation model, 45
expectations, 2, 3, 5, 10, 15, 18, 20, 22, 38, 40, 41, 45, 46, 47, 48, 58, 64, 91, 110, 113, 115, 125, 128, 133, 140, 142
expectations of service, 3
expected and perceived service, 49
experience, 3, 15, 30, 35, 42, 49, 50, 55, 56, 71, 74, 80, 81, 89, 106, 110, 111, 112, 114, 118, 119, 125, 147, 153
explicit, 15, 17, 19, 100, 106
explicit and implicit rules, 15
explicit codes of behaviour, 106
exploratory factor analysis (EFA), 75, 76
exports via, 2
expressing emotions, 37
expression of emotion, 35
expressions or community traditions, 14
expressiveness, 31, 36

F

face-to-face encounters, 44
face-to-face interactions, 3
facial expression, 31
facial gestures, 22
factor analysis, 67, 74, 75, 76, 77, 79, 80, 130, 134, 138, 148
factorability of variable, 75

family and the community above individual interests,, 35
family relationships, 16
favourable policies, 3
favourable word of mouth recommendation, 125
feelings, 31, 33, 34, 37, 102, 103, 106
filial piety, 104, 136
foreign investment, 3
formal and informal behaviour, 37
formal behaviour, 99
formal etiquette, 99, 115, 116
formalisation systems, 101
former colony, vii, 9
forthrightness, 34
France, vii, viii, 4, 5, 6, 7, 8, 9, 18, 32, 66, 100, 109, 112, 115, 117, 119, 129, 132, 135, 137, 142, 148, 150, 153
Francophone Community, vii, 9
free choice, 29
freedom, 19, 29, 37, 83, 84, 115
freedom of press, 115
freedom of speech, 115
freedom of worship, 115
French culture, 71, 91, 99, 103, 104, 115, 120
French economy, 6
French images, 6
French language teaching, 6
French legacy, vii, 9
French long-haul travellers, 8
French society, 99, 101, 103, 105, 108, 116, 117, 119
French tourist market, 7, 11, 90, 110
French visitors, vii, 9, 116, 120
French-style architecture, vii, 9
friction, 34
friendship, 29, 35, 107
front-line employees, 3
fruitful relationships, 42
functional quality, 51

G

gaining a competitive advantage, ix
gateway for other European travelers, 7
generalizability, 126
gestures, 22, 31, 105
gift giving, 31
globally, 1
gradual relaxation of visa regulations, 4
Great Britain, 130, 132
group tours, 4
Guanxi, 104, 105, 107
guest inference about host personalities, abilities and attitudes, 53
guest service evaluations, 125
guest standards of behaviour and needs, 120
guests, ix, 36, 40, 42, 53, 58, 63, 66, 79, 81, 90, 94, 95, 97, 109, 110, 111, 113, 114, 116, 117, 118, 119, 120, 121, 123, 125, 144, 147
guidelines, 20, 101, 124

H

habits, 2, 14
Hampden-Turner and Trompenaars, 23, 30, 117
hardships, 100
harmonious interpersonal relations, 37
harmonious relations, 34
harmony, 25, 29, 33, 34, 37, 83, 84, 86, 99, 102, 106
harmony with nature, 33, 37
hierarchical authority, 100, 115
hierarchical cognitive structure, 23
hierarchical ordering, 25
hierarchical relationship systems, 101
hierarchical society, 99
hierarchical theory of motivations, 19
hierarchical-type terms, 116
hierarchy, 27, 31, 35, 85, 87, 100, 116, 119

high power distance, 33
high quality, 5, 9, 49, 68, 114
high quality French cuisine, 9
high reliability and validity, 73
high uncertainty avoidance cultures, 120
high-context culture, 37
Hinduism, 33
history, 2, 3, 15, 18
Hofstede, 16, 21, 22, 26, 27, 28, 30, 31, 32, 101, 102, 108, 135, 136
holidaymakers, 2, 4
homogeneous sample populations from Western countries, viii, 10
Hong Kong, 129, 136, 139
horizontal networking relationships, 3
hospitality, 13
host, viii, ix, 10, 13, 17, 18, 32, 42, 44, 45, 58, 69, 110, 124
host and guest cultures, viii
host attributes, 50, 51, 53
host perceptions of tourists, 40, 54, 57, 58
host population, 18
host products and services, 56, 58
host-guest mutal perceptions, 53
host-guest service encounter interactions, 58, 124
host-guest service interactions, 45, 58
host-guest social, 55
hosts and guests, 10, 44, 45, 58, 77, 110, 115, 123, 124
hosts-guests, 11, 13, 125
hotels, 3, 68, 70, 125, 126
human, 3, 13, 15, 17, 19, 21, 22, 23, 26, 27, 30, 34, 43, 44, 69, 103, 106
human actions, 15
human behaviour, 19, 22, 26, 31, 147
human emotions, 103
human interactions, 43, 54
human life, 13, 27
human nature, 22, 26
human needs, 21
human personality, 112
human race, 17
human relationships, 30
human resource, 3, 6, 129
human resources, 3
human resources training, 6
human values, 23, 69
humanism, 32
humanitarianism, 37
human-to-human coordination, 43
humility, 20, 103

I

ideal, 14, 19, 108
ideational form of culture, 14
ideological, 14, 15, 28, 150
ideological entity, 14
ideological statement, 28
implicit, 17, 19, 100
implicitly or explicitly acknowledge, 41
implicitly structured activities, 108
improved detection of negative perceptions, 59
inbound tourism, 4
independence, 29, 88, 99
independent, iv, 4, 26, 29, 35, 37, 67, 72, 73, 80, 108, 118, 133, 136
independent travellers, 4, 67
independent variable, 26
indirect culture, 35
indirect in behaviour, 34, 106
indirectness, 35
individual initiative, 108
individual needs, 88, 108, 125
individual obligations to society, 113
individual perception, 30
individual perceptions, 30, 39
individualism, 26, 30, 35, 37, 147, 149
individualism-collectivism, 26, 147, 149
individualist cultures, 102
individualistic and vertical cultures, 117

individualistic cultures, 31
individuality, 102, 107, 108
individuality and freedom, 115
individualized attention, 91, 93, 95, 96, 97, 120
individuals, 19, 20, 23, 26, 30, 31, 34, 36, 37, 44, 105, 108
Indo-China, 2
industrial culture, 17
industry, 2, 3, 5, 17, 43, 124
inferences about another person based on visible or audible behaviour, 53
informality, 36, 100, 103
informality and social reciprocities, 103
infrastructure, 5
inner-directed Western society, 102
inseparability, 50
institutions, 22
instrumental values, 28, 29
insulting, 36
intangibility, 50
interaction behaviour, 41
interaction between a customer and a service provider, 51
interaction difficulties, 41
interactional behaviours, 16
interactional process, 42
interactive dimension of service quality, 51
interactive processes, 51
intercultural communication, 14, 134, 138, 140, 144, 146
intercultural competence, 90, 91, 92, 94, 96, 111, 119, 120, 140
interdependence, 104
interest in others, 85
intermediaries, 35
international commodity exchange, 4
international gateways, 1
international hotel, 3
International Organisation of La Francophonie (OIF), 7
international pattern, 33
international relations, 4
international service providers across Southeast Asian, 125
international tourists, viii, 4, 67, 124, 125, 126
inter-organizational service encounter, 42
inter-perceptions, 13, 54, 124
interpersonal, 25, 27, 28, 31, 33, 34, 35, 42, 43, 69, 70, 102, 107, 110, 112, 128, 137, 140, 142, 149
interpersonal communication, 31, 140
interpersonal encounters, 110
interpersonal exchange, 110
interpersonal interaction and communication, 113
interpersonal interactions, 25, 35, 69
interpersonal relations, 31, 33, 34, 36, 37, 70, 102, 107, 112, 149
interpersonal relationships, 33, 34
interpersonal skills, 43
intimacy, 29, 31
intra-group social harmony, 100
intra-organizational, 42
intrapersonal, 28
intra-regional tourism strategy, 1
introduction of Doi Moi, 3
irritating, 36
issues, 19, 86, 106, 107

J

joint-venture enterprises, 2

K

Kaiser-Meyer-Olkin (KMO), 75, 76, 80, 84, 87, 88, 89, 91, 92, 93, 94, 96, 97
kin group, 99, 100, 115
Kinh trong, 106
kinship, 101

knowledge and self-actualization functions, 19
Korea, 35
Kroeber and Kluckhohn, 14, 16, 17, 21

L

labour market, 3
lack of knowledge, 3
Landis and Brislin, 16
language habits, 14
large influx of tourists, 5
Latin countries, 32
Latin oriented cultures, 105
laws, 100
Le phep, 106
learning, 13, 26, 37
legacy of grand culture, 2
leisure, 3
less materialistic and receptive to nature, 35
level of performance, 46
linguistic communication, 43
logic of comparative analysis., 62
low power distance, 33
loyalty, 20, 31, 37

M

maintain harmony in human relations, 34, 106
maintain social harmony, 34, 37, 86
Malaysia, 131
management, 2, 3, 5, 14, 17, 49, 52, 70, 128, 129, 134, 135, 136, 137, 138, 139, 141, 143, 145, 146, 148, 150, 151, 153
managing conflict styles, 34
market economy, 3, 5
market economy tourism sector, 5
market potential, 1, 2
market segment, 26, 124
market segmentation, 26, 124

marketing, 4, 13, 14, 42, 45, 48, 64, 74, 124, 125, 126, 128, 130, 131, 132, 133, 134, 135, 139, 140, 141, 142, 143, 144, 145, 147, 148, 150, 153
marketing practices, 124, 143
masculinity-femininity, 26
Maslow, 19, 24, 141
material and non-material elements, 13
material aspects, 14
material culture, 14
material elements, 15
material manifestation, 13, 14
material one, 113
material well-being, 37
means and ends, 19
measure the attitudes and opinions, 64
measurement, 27, 28, 69
measurement techniques, 27, 28, 61
Mekong sub-region, 1
mental programming, 15, 20, 32
mental programs, 24
meta-perceptions, 54
methodological problems, 61
Middle-Eastern culture, 22
Min, Ho Chi, 68, 131
misinterpretation, 32
misunderstanding, 36, 41
misunderstandings of interpretation, 16
modes of behaviour, 23, 28
moral and social justification, 108
moral obligations, 104
morality, 14, 28, 33, 37, 104
motivational value, 26
multicultural clientele, 124
multicultural needs and wants, 125
multidimensional, 14
multi-item measurement, 73
mutual obligations and reciprocation, 36
mutual perceptions, viii, 10, 11, 123
mutual togetherness, 108

N

national cultural characteristics, 41, 58
national culture, 17
national development, 2
nationality, 109
nation-state, 17
natural, 1, 2, 118, 132
negative emotions, 34
negative impressions, 40
negative opinions, 32, 34
negative perceptions, viii, 10, 41, 53, 54, 56, 57, 58, 124
negative tourist perceptions, 52
neutral cultures, 106
Nha nhan, 106
non-action, 113
non-assertive, 102
non-material, 14, 37
non-materialistic orientation, 37
non-verbal behaviour, 44, 86
non-verbal communication, 31, 32, 105, 106
non-verbal cues, 105
normative values, 24
norms, 14, 16, 18, 20, 21, 22, 23, 24, 31, 39, 126
nostalgic appeal, vii, 9

O

oblique rotation, rotated factors, 77
occidental urban middle class explorers, 9
official development assistance (ODA), 6, 7
opportunities, 2, 5, 105
orientations-conceptions, 22
orthogonal rotation, 77, 80
outbound market for expenditure, 7
outer-directed Asian societies, 102
overall holiday satisfaction, viii, 45, 55
overall perception of tourism products, 55
overall satisfaction, 69, 71, 125, 126
overall tourist satisfaction, 43

P

parent culture, 18
particularistic cultures, 109
patterned material for perception, 39
patterns of behaviour, 13
patterns of culture, 15
patterns of social interaction, 32
patterns of verbal, 31, 32
people perceptions, 53
people status, 101
people-oriented, 35
perceived cultural dissimilarity, 54, 56
perceived exoticism, 109
perceived service quality, 49
perceptions, ix, 5, 10, 14, 16, 17, 21, 24, 25, 26, 28, 32, 38, 39, 41, 42, 43, 44, 45, 46, 51, 52, 53, 54, 55, 56, 57, 58, 63, 69, 70, 71, 90, 91, 95, 110, 111, 118, 119, 120, 123, 124, 125, 130, 131, 133, 137, 142, 145, 146, 147
perceptions of one's own, 54
perceptions of other people, 54
perceptual differences, 44
perceptual mismatching, 54
performance quality, 48, 52
personal achievement, 118
personal domain, 14
personal freedom, 37
personal identity, 108
personal interests, 102, 104
personal or social values, 28
personal relations, 107
personal relationship, 107
personal value, 22, 23, 27, 39, 138
personal values, 22, 23
personality, 108
personalized attention, 91, 92, 95, 97, 111, 119, 120

Index

person-to-person tourist interactions, 43
physical environment, 23
physical quality, 51
physical resources, 42
physically attractive, 56
physiology, 38
pilgrimage to the erstwhile theatres of pitched battles, 9
polite, 20, 29, 83, 84, 90, 91, 92, 94, 95, 96, 97, 101, 102, 103, 106, 115, 116
politeness, 101
political activities, 14
political or religious ideology, 20
political system, 101
political systems, 101
politics, 36, 107
position, 2, 4, 50, 81, 100, 101, 116
positive affiliation, 56
positive or negative disconfirmation, 48
positive perceptions, 51, 55, 56, 57
positive perceptions towards hosts, 51, 55
positive service perceptions, 52
potential tourist resources, 2
power-distance, 26
powerful motivations, 2
pragmatism, 32
predictability, 101
preferable existence, 21
preserving its national heritage, 5
prestige, 102
prevailing behaviour, 99, 116
primary values, 27
principal components analysis (PCA), 75, 76, 77, 79, 80, 81, 89, 95, 120, 121
principles, 19, 23, 69, 85, 99, 100, 108
privacy, 57, 86, 87, 100, 103, 107, 119, 120, 125
privatization, 2
probability rather than non-probability sampling design, 66
procedural and convivial dimensions, 51
production and distribution of services, 118
productivity, 43
products and services offered and consumed, 124
professional and social interactions, 44
professional interaction, 5, 44, 45, 55, 58, 110
professionalise, 2
professionals, 42
programming, 15, 20, 32
progress-oriented, 35
project, 101
promotion, 124, 128
proper treatment of guests, 42
prospects of repeat visitation, viii
protection, 29, 102, 107
Protestant values, 33
Protestantism, 33
psychological concepts, 16
psychology, 14, 38, 53, 128, 129, 130, 131, 133, 136, 137, 138, 140, 141, 143, 147, 151
public and private cultures, 17
purchasing power parity, 6

Q

qualifications, 101
qualified staff, 5
qualitative and quantitative approaches, 64
qualitative data, 64, 65
quality conscious and culture loving, vii, 8
quality of opportunity, 52
quality of service attributes, 51
quantitative and qualitative approaches, 64
questioning, 31, 34, 106
questionnaire, 69, 70

R

race, 17, 18
random sampling, 65, 67

rationalism, 32
reality, 2, 14, 18, 25
recognition, 29, 100
recommendations, 124
regional subcultures, 18
regionally, 1
regulations, 4, 100
relationship between culture and perceptions, 39, 41
relaxation, 4
reliability, 46, 64, 70, 72, 73, 74, 133, 143
reliability assessment, 73, 133
reliable and valid measurements, 62
religion, 15, 18, 36, 44
religious beliefs, 108
religious philosophies, 33, 35
Renqing, 104
Renqing and Bao, 104
representativeness, 126
research instrument, 61
researchers, 22, 27, 43, 45, 110
reserved, iv, 36, 37
resolution of conflict, 19
resources, 2, 5, 14, 42
resources training and management, 5
respect, 20, 22, 29, 33, 35, 85, 86, 87, 100, 101, 102, 103, 105, 107, 114, 116, 119, 125
respect and deference, 119
respect privacy, 107
respond to the needs of the French tourists, 10
responding better to the diverse cultural needs of tourists, 59
responsiveness, 46, 70, 90, 92, 94, 95, 96, 111
restaurants, 68, 70, 125, 126
resurgence with the renovation of colonial-style properties and restaurants, vii, 9
Rokeach's Value Survey (RVS), 28, 69
role performance, 43
Roman language, 33
rules, viii, 3, 10, 15, 16, 17, 18, 21, 22, 24, 25, 26, 30, 31, 36, 42, 43, 44, 45, 69, 70, 85, 99, 100, 101, 102, 109, 123, 124, 126
rules of behaviour, viii, 3, 10, 13, 17, 43, 44, 45, 53, 63, 69, 75, 78, 79, 80, 81, 85, 95, 99, 102, 109, 112, 120, 123, 124, 125, 126
rules of etiquette, 70, 85, 101
rules of formal behaviour, 116
rules of formal etiquette, 99, 116
rules of interactions, 42
rules of social interactions, 44, 123

S

salient attributes, 53, 59
salvation, 19, 29, 82, 83, 84, 125
sampling, 63, 65, 75, 76, 80, 81, 83, 84, 88, 89, 91, 93, 94, 97, 148
sampling adequacy (MSA), 75, 76, 80, 81, 83, 84, 88, 89, 91, 93, 94, 97
sanctions, 20, 24
satisfaction, ix, 5, 10, 11, 15, 16, 17, 32, 40, 43, 44, 45, 48, 49, 54, 55, 63, 66, 69, 70, 71, 74, 75, 77, 79, 80, 81, 89, 93, 94, 95, 96, 110, 112, 114, 118, 120, 123, 124, 125, 142, 143, 144, 146, 149, 150
satisfaction with actual experiences, 112
satisfaction with actual tourism services consumed or offered, 112
save face, 34
saving face, 31, 37
second leading bilateral donor, 6
secondary values, 27
security, 7, 19, 26, 29, 44, 70, 82, 84
self-actualization, 19
self-confident, 108
self-control, 105
self-description, 28
self-discipline, 29
self-discipline and their behaviour, 118

self-effacement, 34
self-esteem, 25, 29, 34, 37, 108, 120
self-expression, 33
self-fulfilling, 108
self-improvement, 107
self-interest, 108
self-interested, 108, 120
selfishness, 37
self-motivated, 37, 118
self-presentation, 31
self-reliance, 99
self-reliant, 29, 108, 120
self-restraint, 34
self-sufficient, 29, 35
self-worth, 108
semi-interview and personal observation, 65
senior market, 9
sense of belonging to groups, 37
sense of shame, 31, 70, 85, 102
service attributes and performance, ix, 11, 45, 53, 55, 58, 70, 71, 79, 80, 89, 92, 110, 112, 119
service delivery, 50, 51, 52, 55, 69, 118
service encounter, ix, 10, 13, 16, 41, 42, 43, 44, 45, 51, 53, 58, 106, 109, 115, 123, 124, 125, 126, 147, 148
service encounter interactions, 16, 58, 123, 125
service industries, 42
service interactions, 10, 44
service personnel, 41, 42
service provider, 3, 5, 10, 17, 24, 42, 43, 45, 70, 124, 125
service provider attributes and performance, 55, 59, 81
service providers, viii, 42, 43, 51, 52, 53, 54, 56, 70, 79, 81, 110, 118, 119, 124, 125
service quality, ix, 5, 42, 43, 45, 46, 47, 48, 49, 50, 51, 52, 54, 55, 58, 68, 70, 110, 116, 118, 125, 126, 128, 130, 131, 133, 134, 136, 139, 140, 142, 143, 148

service quality construct, 47
service quality perceptions, 125, 126, 128
services, 3, 42, 43, 70, 110, 123
services deliver performances, 50
SERVPERF, 45
SERVQUAL, viii, 45, 46, 47, 70, 128, 130, 142
SERVQUAL instrument, 46, 47, 70
set of values, 18
Shintoism, 33
showing, 20, 31, 37, 85, 86, 88, 101, 103, 104, 105
similarities, 16, 17, 38, 39, 62, 69, 81, 106, 118, 119, 124
skilled labour, 5
smart, 33, 90, 92, 94, 96, 116, 147
sociability, 108
social action and productive activity, 63
social activities, 43, 44
social acts, 14
social and family harmony, 99
social anthropologists, 29
social behaviour, 17, 20, 21, 31, 100, 102
social categories, 31
social class, 15, 102
social cognition, 24
social cognitions, 24
social context, 70
social disruption, 34
social harmony, 86, 87, 102, 113
social harmony in interpersonal relations, 113
social interaction, viii, 13, 18, 25, 30, 31, 37, 41, 44, 54, 65, 70, 109, 113, 123, 124, 133
social interaction in the intercultural service encounters, viii
social interactions, 13, 18, 31, 44, 54, 65, 70, 109, 113, 123, 124
social issues, 19, 145, 147, 151
social life, 19, 30, 106
social network, 104, 105

social obligations, 33
social order, 99
social organization, 22, 102
social position, 99, 100, 115, 116, 119
social processes, 124
social recognition, 29, 83, 84, 100, 120
social relations, 2, 54, 85, 87, 101, 102, 103, 105, 128
social relationships, 2, 103
social rules, 70
social situations, 44, 69, 109
social stability, 102
social standing, 100, 114
social stratification, 85, 100, 108, 115
social values, 19, 62, 137
societal culture, 17
societal level, 20
societal/national culture, 17
society, 15, 17, 18, 20, 22, 23, 25, 27, 30, 35, 37, 43, 99, 101, 103, 105, 108
society-member relationship, 30
society-oriented, 35
socio-demographic backgrounds, 66
sociology, 14, 127, 134, 137, 143, 148, 150
solution, 22, 85, 86
South East Asia, 1, 4
South East Asia's tourism development trends, 4
Southeast Asia, 5, 125, 145
Spain, 33
speaking distance, 36
special interest holidays, 8
spirit of Buddha, 100
spiritual, 14, 29, 103, 113
spiritual life, 113
spirituality above material well-being, 35
stable tourism, 3
standalone destination, 4
standard in service quality, 49
standardized criteria, 50
statistical information on visitor numbers, 5

status, 31, 70, 86, 87, 100, 101, 103, 105, 115, 117, 119
stereotype, 40
strategic partnership, 6, 142
stratification, 85, 100, 108
stratification criteria, 66
stratified convenience random sampling, 66
strong propensity to travel abroad, 8
subcultures, 13, 17, 18
subjective culture, 38, 149
subjective customer perception of service, 50
subjective perceptions, 43
successful interaction, 20
survey and personal face-to-face interviews, 68
survey sample methodology, 64
symbolic statement, 14
synchronization, 117

T

tangible products, 50
tangibles, 16, 46, 70
Taoism, 33
target culture, 56
taxing and demoralizing, 34
technical quality, 51
techniques, 28, 31
Ten Commandments, 33
terminal values, 25, 28, 29
tertiary values, 27
the five senses, 38
thoughts, 16, 20, 33
time as monochrome, 117
top-down, 3
tourism, ix, 1, 2, 3, 5, 10, 43, 44, 45, 70, 123, 124, 126
tourism development process, viii
tourism development trends, 1
tourism education, 3

tourism growth, 2, 4
tourism marketing, 53, 59, 124, 132, 140, 145
tourism marketing planning, 59
tourism marketing strategies, 124
tourism products, 2, 4, 77, 123
tourism products and services, 4, 77, 123
tourism promotional strategies, 11
tourism service encounters, 18
tourism service providers, 3, 5, 124
tourism source market, vii, 9
tourist generating countries, 124
tourist overall holiday experience and satisfaction, 58
tourist perspective, 39, 55, 120
tourist psychological needs and experiences., 124
tourist vacation choice decisions, 58
tourist–host mutual perceptions, viii, 18
tourists and service providers, viii, 5, 10, 17, 58, 65, 118
tourists/hosts encounters, 42
tradition of ethnic groups, 2
traditional ways of behaviour, 99, 116
traditions, 13, 14, 15
traditions and resources, 14
training, 3, 5, 124
transactions, 70, 107
transcendental level, 24
transforming economic, 2
travel long-haul, vii, 8
Triandis, 16, 19, 27, 30, 31, 38, 73, 127, 136, 149, 151
Trompenaars, 23, 30, 31, 101, 102, 105, 106, 135, 149
trustworthiness, 90, 92, 96, 111
truthfulness, 34, 113

U

uncertainty avoidance, 33, 100, 108

underdeveloped countries, 2
underlying belief, 19
unfriendly attitudes, 41
unique customs and habits, 2
United Nations Security Council, 5
United States, 135, 137, 142, 150
univariate descriptive statistics, 72
universal (human) culture, 17
universal or culture-free theories and concepts, 62
Universalism-Particularism dimension, 109
universalistic cultures, 109
unpleasant encounters, 41
unsatisfactory experiences, 10
unspoiled natural environments, 1
urban, 2
urban tourism, 2
USA, 18, 109, 133

V

vacation choice, 41, 147
value and attitude, 25
value orientation, 20, 22, 24, 26, 27, 30, 32, 138
value-expressive, 25
values, viii, 3, 10, 13, 14, 15, 16, 17, 18, 19, 20, 21, 23, 24, 25, 26, 27, 28, 30, 32, 33, 35, 37, 38, 39, 42, 43, 44, 54, 56, 61, 62, 68, 69, 72, 80, 83, 84, 106, 108, 109, 123, 124, 125, 126, 127, 128, 129, 131, 133, 135, 137, 138, 139, 140, 141, 146, 147, 149, 151
values are patterns of choice, 21
values are symbolic statements, 14
values-attitudes-behaviour, 24
variables, 19, 23, 30, 85, 86, 88, 89, 123
variations, 22, 44, 70
verbal and nonverbal means, 39
verbal codes, 37

verbal communication, 91, 92, 95, 97, 111, 119
verbal complements, 22
vertical organization structure, 3
vertical social hierarchy, 114
Vietnam, 1, 2, 3, 4, 5, 10, 18, 35, 45, 72, 101, 102, 105, 107, 109, 123, 124, 125, 131, 137, 149
Vietnam as a holiday destination, viii
Vietnam inbound tourism, 1
Vietnamese, viii, ix, 2, 3, 4, 6, 9, 10, 14, 16, 23, 26, 29, 33, 34, 45, 53, 54, 58, 63, 65, 66, 67, 68, 69, 70, 71, 72, 75, 76, 77, 79, 80, 81, 82, 84, 85, 87, 88, 89, 92, 93, 94, 95, 96, 97, 99, 100, 101, 102, 103, 104, 106, 107, 108, 109, 110, 111, 112, 113, 114, 115, 116, 117, 118, 119, 120, 121, 123, 124, 126
Vietnamese culture, 14, 33, 70, 105, 111, 116, 117, 120
Vietnamese managers, 2
Vietnamese service providers, viii, 10, 24, 45, 54, 59, 66, 71, 80, 81, 110, 113, 119
Vietnamese society, 99, 100, 104, 107, 108, 113, 114, 116, 119, 120
Vietnamese tourism sector, 4
violence, 33

vocabulary, 36
vocational training, 3
volume of interaction, 31

W

ways of life, 35
welfare, 29, 37
well-being, 35, 37
well-mannered, 29, 102
West, 30, 35, 36, 101, 131, 149
Western and Asian language groups, 66
Western and Eastern societies, 33, 34, 35, 37, 41
Western countries, 3, 106
Western Europe, 33
Western friendliness and informality, 36
Western management practices, 2
Western markets, 4
Western societies, 32, 34, 35, 36, 37, 99, 103, 105, 112, 115, 117
wide cross-section of travellers, 2
world travel awards, 4
Wu Lun, 102